ECOLOGICAL RESEARCH
WITH CHILDREN AND FAMILIES

From Concepts to Methodology

Alan R. Pence
EDITOR

Foreword by Urie Bronfenbrenner

Teachers College, Columbia University
New York and London

ACKNOWLEDGMENTS

This publication would not have been possible without the financial support of the Social Sciences and Humanities Research Council of Canada (SSHRC). The Council's support for an intensive three-day workshop on the theme of ecological research with children and families served as the forerunner for this publication. The assistance of Sandra Griffin and Reed Early at the Victoria workshop is also gratefully acknowledged.

Published by Teachers College Press, 1234 Amsterdam Avenue, New York, NY 10027

Library of Congress Cataloging-in-Publication Data

Ecological research with children and families : From concepts to methodology
/ Alan R. Pence, editor: foreword by Urie Bronfenbrenner.
   p. cm.
Bibliography: p.
Includes index.
ISBN 0-8077-2914-0. ISBN 0-8077-2913-2 (pbk.)
   1. Child development—Congresses. 2. Family—Congresses.
3. Environment and children—Congresses. 4. Child care—Congresses.
5. Human ecology—Research—Congresses. I. Pence, Alan R.,

HQ767.9.E34 1988   88-4814
155.4—dc19   CIP

Manufactured in the United States of America

93 92 91 90 89 88   1 2 3 4 5 6

*To Orville, Evelyn, Don, Jean, Nathan, and Ruth,
those grandparents who have shaped my own understanding of
family and provided my children with the gift of generations.*

CONTENTS

FOREWORD

URIE BRONFENBRENNER

This is a foreword to a fascinating book. Edited by Alan Pence, the book is fascinating not only for the new knowledge and the new ideas that it presents, but perhaps even more for its documentation of a developmental life course, now at mid-passage. The life course is not that of an individual, but of a scientific undertaking—an effort to implement a new, or to speak more accurately, newly-rediscovered research paradigm for studying human development. Called the "ecology of human development," or more succinctly, "development in context," the model was first explicitly proposed in the middle and late 1970s (Bronfenbrenner, 1974, 1975, 1977a, 1977b, 1979). Although I am sometimes identified in the pages that follow as the originator of this perspective, that designation claims too much. Perhaps the most that can be said is that mine was the first attempt to give the paradigm systematic form; thus, my role was more that of a scientist-historian who discerned in the work done by different investigators, in different branches of the field, certain common and complementary features that pointed to an emerging new conceptual model. In the effort to make these features explicit, and to link them together in a comprehensive frame, I have also introduced some new theoretical concepts of structure and process.

It is from this viewpoint of a partial outsider that I find the contributions in this volume so intriguing, for they tell us what can happen when a newly-defined theoretical orientation is taken seriously and translated into action—in this instance, not only in science, but also in policy and practice.

It is also from the perspective of a partial outsider that I have been invited to write the foreword to this volume, and to highlight what is in store for the reader. But here a caveat is in order: I am only a *partial* outsider—one who is far from *impartial*. One can hardly claim to have no personal involvement in the subsequent course of an undertaking that one has helped to set in motion. As witness to that fact, I could not read the pages you will be reading without experiencing feelings of both pride and pain. The pride was in what others had accomplished with ideas that I had helped to formulate; the pain came from those of my ideas that I had failed to formulate clearly enough so that they could be understood and carried out well, or —perhaps worse yet—ideas that were formulated *too* clearly, so that they were carried out *too* well.

I recognize, somewhat reluctantly, that I owe it to the reader to tell about both kinds of ideas. To start on the happier note: A decade ago, one of the first salvos to be fired by proponents of the ecological approach against the then-prevailing modes of developmental research was the charge that investigators were spending most of their time studying "they strange behavior of children in strange situations with strange adults for the briefest possible periods of time" (Bronfenbrenner, 1974). In short, *development was being studied out of context*. As the chapters in this book dramatically demonstrate, that is no longer the case today. Not only is development being observed in the actual settings where human beings live and grow, but the resultant findings provide considerable evidence validating a central thesis of the ecological approach: that the conditions under which human beings live have a powerful effect on how they develop. In particular, the results reveal the power of what Robert Glossop, in his penetrating introductory chapter, identifies as "the cornerstone of the ecological frame of reference." That is, the proposition that the developmental processes taking place in the immediate settings in which human beings live, such as family, school, peer group, and workplace, are profoundly affected by conditions and events in the broader contexts in which these settings are embedded.

For example, in an analysis of the actual processes occurring in different types of day care settings, the work of Alan Pence and Hillel Goelman offers evidence that such processes appear to account for greater variation in the development of young children who experience family day care than of those enrolled in center care. Although the author is appropriately cautious in suggesting explanations for this contrast, one possibility may be the greater variability in the quality of care found among family day care homes as compared to similar environments observed across center settings. Equally relevant, and perhaps more troubling, is the finding that disproportionate numbers of children from low-resource families were also entering low-resource day care environments. As Goelman notes, what the data reveal is "a portrait of an unsupported ecology of family life." Here, science provides validation from

the lives of today's children of an ancient Biblical warning attributed to Jesus: "For whosoever hath, to him shall be given, and he
shall have abundance; but whosoever hath not, from him shall be
taken away even that he hath" (Matthew 13:12, KJV).

This so-called "Matthew Principle" is accorded scientific legitimacy by research findings reported in several chapters of this volume, among them Johnson and Abramovitch's analysis of the
disruptive effects of paternal unemployment on family processes, and
Michael Lamb's systematic demonstration of the "unsupported
ecology" of teenage parents.

But science does not necessarily follow Biblical warnings of ret-
ribution. The research findings cited above represent documenta-
tions of existing reality. The ecology of human development,
however, is not just a scientific statement of the status quo; rather,
it is conceived as a system of organism–environment interrelationships that exhibit both stability and susceptibility to change.

The sources of change are twofold: The first is intervention from
the outside; the second, of equal importance and necessity, is the
initiative taken by an active living organism whose basic impulses
are directed toward survival, constructive action, and psychological
growth. It is the complementary conjunction of the two vectors that
drives development. Demonstration of these forward thrusts is also
documented in the chapters ahead. Witness, for example, Lamb's
finding, among unwed teen-age parents, of "a striking willingness
on the part of the women not to seek elective abortions and on the
part of their partners to assume some responsibility for their partners and children." At the same time, measures of paternal involvement were found to be related to the availability of social support.

The same convergence of positive vectors from within and
without appears in Moncrieff Cochran's account of processes and
outcomes in a community-based intervention program designed to
enhance family empowerment through building on family strengths
and fostering supportive networks. The program showed its strong-
est effects on single parents and their preschool children, with the
mothers' self perceptions, social networks, and joint activities with
the child serving as powerful mediating variables.

Throughout the volume, the contributions of the ecological perspective are demonstrated not only in the authors' scientific findings but also in the design both of their research, and of their
programs and policies. In the second domain, Donna Lero illus-
trates possibilities for the more creative use of national surveys as
tools for understanding the ecology of the child. In a still broader

frame, Wes Shera provides an intriguing account of the explicit use of an ecological model in the design, creation, and evaluation of Tumbler Ridge, a new coal mining community in northern Canada. Among the criteria of evaluation were residents' ratings, including those made by elementary school children, of quite concrete, down-to-earth experiences in the everyday life of the town: the level of neighborhood grocery prices, housing choice, health and social services, day care facilities, schools, library services, activities for youth, municipal services, and the quality of the community as a whole. Shera's design implements the nested ecological model at every level from the micro to the macro. Shera then applies what Brian Wharf, in a final chapter on implementing the ecological per-spective in policy and practice, astutely designates as the ultimate test of an ecological environment for human development—that its resources "must be delivered within pram-pushing distance of the individual." Testimony that Tumbler Ridge passed this exacting re-quirement comes from the group that is perhaps in the best posi-tion to judge: Among those giving the community its highest marks were the schoolchildren. As I have suggested elsewhere (Bronfen-brenner, 1970), the most telling criterion for evaluating the health of a society is "the concern of one generation for the next" (p. 1).

One does not necessarily have to start from scratch. Following an incisive analysis of what he characterizes as the "residual ap-proach to social policy" prevailing in contemporary America, Wharf draws on research findings from a range of successful programs in order to delineate the complex of ecological elements that make a difference. Three of these critical features are brought to the fore in separate chapters by James Anglin and Douglas Powell. First, like Cochran, these authors emphasize the importance, even the neces-sity, of the empowerment of those who are the intended benefici-aries of policy and practice. Indeed, in the last analysis, they turn out to be the principal agents of change.

A second essential feature stressed by Anglin and Powell, and richly illustrated from their own research projects, is the impor-tance—both prior to and during the execution of the program—of discovering and, to the extent possible, responding to the differing characteristics, needs, and initiatives of program recipients. To state the issue in theoretical terms, the program itself must be viewed as yet another nested, dynamic subsystem within the family's ecology. Powell points out that the resultant awareness of variations among recipients can and should lead to a recognition of the limitations of any program's capacity to be responsive to the diversity of people's

characteristics and needs. Anglin takes the implication a step further in arguing that a genuine commitment to responsiveness implies a collaborative working relationship in which the recipients
themselves participate in the direction and development of the program. Thus the program itself provides yet another arena for self-
empowerment.

The most significant contribution of these two chapters, however, is one that the authors do not state explicitly, but which they
communicate even more eloquently through the analyses that they
carry out. In conceptualizing and examining an action program as a
subsystem in a larger ecology of human development, they also begin to view the program subsystem as itself undergoing development in context, and to treat it accordingly. Thus the program
becomes, in the researcher's perceptions—and, thereby, in reality—a social organism that, over time, exhibits greater differentia-
tion as it actively, and sometimes painfully, accommodates to the
evolving initiatives, responses, and life circumstances of the families it serves.

A third critical element emphasized in these two (and indeed
in all of the) chapters of the volume is embodied in what I once
identified as the "only proposition in social science that approaches
the status of an immutable law" (Bronfenbrenner 1979, p. 23). I refer to W.I. Thomas's inexorable dictum of more than half a century
ago: "If men perceive situations as real, they are real in their consequences." (Thomas & Thomas, 1927, p. 572). Anglin, in particular, traces the progressive reformulation, in his own project, of ends
and means as the program moved away from an objective toward a
phenomenological orientation. The shift leads him to a position in
which "the distinction between research and intervention becomes
no longer tenable," "the separation between researcher and subject
begins to dissolve," and the "qualitative review of concepts and
themes replaces the quantitative analysis of questionnaire data as the
primary source of understanding." Moreover, the theoretical basis
for this view, and the ground for its proposed logical and experiential necessity is found in the writings of none other than the author
of these lines.

I cannot deny the responsibility implied in this gracious attribution by Anglin, for he cites incontrovertible evidence from my
own writings of repeated theoretical arguments, and impassioned
advocacy, for the essentiality and primacy of a phenomenological
approach both in research on development in context, and in the
design and execution of programs and policies based on this ap-

proach. For example, I can only concede that Anglin has counted correctly when he asserts that the necessity to examine "the way the environment is perceived, and the meaning it has for the developing person" is repeated "no less than 20 times" in my formal exposition of the ecological model.

It is with no particular enthusiasm, however, let alone satisfaction, that I acknowledge these points. The reason: When one reviews the research conducted within an ecological framework in the years since the publication of my 1979 monograph on this topic, one of the principal generalizations that emerges does not exactly validate the priority that I accorded to the phenomenological imperative. As I have documented elsewhere (Bronfenbrenner, 1986a–c; in press; Bronfenbrenner & Crouter, 1983), findings to date (now adding those reported in this volume) reveal that the most powerful environmental forces shaping human behavior and development emanate primarily from the objectively-measured, objective conditions and events occurring in the life of the developing person. Examples include the material circumstances in which that person and his or her family live, the actual behaviors of other parties toward the person in question in face-to-face situations, and the actions and decisions taken in external, structurally-defined contexts, such as businesses, government agencies, social organizations, and other institutions in the public and the private sector.

This is not to imply that how these agents and agencies are perceived, and how persons perceive themselves, do not play a significant role in shaping, impeding, or accelerating the course of behavior and development. As almost every chapter in this volume demonstrates, recognition of the belief systems of self and other is crucial both to the scientific understanding and the social harnessing of the forces that shape human development. But the power of belief systems is realized only through their translation into overt behavior, as human beings confront, cope with, alter, and create objective conditions and events. To add a necessary companion principle to Thomas's classic injunction: "Real situations not perceived are also real in their consequences."

This complementary principle has at least two compelling corollaries for research design and method. The first requires that investigators, while making full use of a phenomenological perspective, give equal priority to the recognition and analysis of the objective conditions, events, and processes taking place in the life of the developing person. In the last analysis, it is the interaction between these two domains that shapes both the developmental course and its outcomes.

How does it happen, then, that what I originally set forth as a definitive statement of the ecological model turned out to be so one-sided? To be sure, I made a number of dutiful references to the critical role of objective factors. But, as Glossop, in his "reappreciation" of my 1979 volume points out with this customary perspicacity, my emphasis was strongly on the phenomenological side: "[Bronfenbrenner] argues that the most important elements in the environment that influence human development are those that 'have meaning for the person' and that what is important to understand is the environment 'as it is *perceived* rather than as it may exist in 'objective reality'" (italics in original).

How am I to explain this pronounced one-sidedness in my original presentation? One does not have to look far to detect the roots of the bias, and, again, Glossop has pointed to the ground in which they lie. In tracing the origin of my ideas in the work of others, he correctly identifies their principal source: "Most particularly [Bronfenbrenner] builds upon the phenomenological concepts of the 'life space' and the 'psychological field' as incorporated into the systematic theory of Kurt Lewin.'"

In a comprehensive review of developmental theory that I wrote many years ago (Bronfenbrenner, 1951), at a time when I was studying Lewin's work most intensively, there appears the following summary statement: "For Lewin what is most relevant in the environment is not what is objectively there . . . but what is perceived." I then had added, in a footnote, that, while Lewin acknowledged that "the objective reality is nevertheless important," he gave it short shrift, referring to it only parenthetically as the "foreign hull." The final sentence in the footnote reads: "Lewin is somewhat unclear and contradictory on this point" (Bronfenbrenner, 1951, p. 211). Now, three-and-a-half decades later, it become apparent that what one inherits from a great mentor is not only his insights, but also his oversights.

With the ever-marvelous wisdom of hindsight, if I were writing my 1979 monograph today, I would not delete a single one of the 20 references to the essentiality of pursuing a phenomenological approach in developmental science and social policy. But then—to right the imbalance—I would add a comparable number of, I hope, similarly persuasive arguments for according equal priority to the analysis of objectively-identified conditions, events, and processes existing in the developing person's objective world.

The second corollary focuses on the even more basic issue of the nature of scientific method as it relates to the nature of the scientist, his strengths and his limitations. (Unfortunately, it is still the case that even in the field of human development, the scientist is usually "he," as with nine out of twelve authors in this volume). As Anglin so movingly documents in tracing the life course of his own research project, the investigator's "values, beliefs, judgments, preferences . . . importantly impact on every aspect of the study." Anglin is correct in regarding this phenomenon as having fundamental importance. In fact, modern science had its beginnings in the effort to eliminate, or at least to minimize, the bias introduced into research results by what was then called the "personal equation." The most effective strategy that scientists then applied to counteract this bias was to introduce "external" procedures that could challenge the bias in ways that were as independent as possible of influence from the investigator's predilections. The strategy is admittedly imperfect, but it remains the most effective one known today.

But today, application of the strategy has become a far more complex and difficult task, and for good reason. Biases cannot be corrected until they are discovered and acknowledged. One of the principal advances achieved by the ecological approach, with its insistence on research in real-life settings and the injunction to understand those settings from the perspective of those who live there, has been to make investigators painfully aware of the fallacious and simplistic nature of their stereotypes, scientific no less than personal. Both the progress and the pain are eloquently conveyed in Anglin's perceptive analysis. But the fact that we are constantly discovering new sources of bias as Anglin has done—bias not only in what the investigator sees, but fails to see—does not mean that the strategy should be abandoned, but rather that it should be pursued even more imaginatively and relentlessly! Indeed, some of the more promising directions for that pursuit are those suggested by Anglin himself, along with other ingenious designs and devices employed by contributors to this volume and by many other investigators seeking to probe the twin realities of human development in context.

Moreover, when viewed from a developmental perspective, the case for confronting objective reality has deeper roots. Science has its origins and derives its power from two extraordinary emergent capacities of the developing human being. The first is imagination: conceiving of the possible and the impossible. The second is reality testing: the capacity, through logical thought and experiment, to distinguish the possible from the impossible in the light of actual conditions, limitations, demands, resources, and opportunities objectively existing not only in the environment, but also within the person.

This brings me to the final, uncomfortable obligation of acknowledging yet another and more extreme one-sidedness in my original exposition of the ecological model, an imbalance that, once again, also characterizes the subsequent work, both basic and applied, that has since been carried out in an ecological perspective. This time, I will resist the temptation to begin by externalizing the problem. Many authors of the chapters in this book, as well as other researchers on development in context, have cited as the point of departure for their work the definition of ecology appearing in my 1979 monograph; namely, "the scientific study of the progressive mutual accommodation between an active, growing human being . . ." (p. 21). I stop there, because nothing more is said about this, the first critical term in the ecological equation; the rest of the definition, taking up all of the next four lines, has to do solely with the environment. Worse yet, the same proportion holds for the monograph as a whole.

A comparable asymmetry characterizes the research produced by adherents of the ecological perspective in the years that followed. The numerous studies of development in context that have been published over the past decade, as well as the chapters in this volume, present much new knowledge about the complex structure of an environment conceived in systems terms and the bidirectional processes operating both within and across its constituent subsystems. In sharp contrast, these studies offer far less new knowledge about the evolving complex structure of the developing person and the bidirectional processes operating both within personal subsystems and within the subsystems linking the developing person to the outside world. As I have written elsewhere: "In place of too much research on development 'out of context,' we now have a surfeit of studies on 'context without development'" (Bronfenbrenner, 1986a, p. 288).

To the extent that my own neglect of the organism in the ecological equation has contributed to this general trend, I must further acknowledge that this error of omission was the result of a deliberate decision. In my 1979 monograph (p. 12), I expressed "the conviction that further advance in the scientific understanding of the basic intrapsychic and interpersonal processes of human development" must wait upon the formulation and implementation of a more differentiated and dynamic conception of the environment. I then addressed myself to the latter task. From one viewpoint, that decision may be seen as an appropriate pragmatic ordering of priorities; from another, as an all-too-familiar device for avoiding a difficult task—putting off till tomorrow what could and should be done today.

There are occasions when even the illuminating wisdom of hindsight does not reveal what should and could have been done yesterday. In fact, this is one instance in which present wisdom offers a clearer view. Today, there can be no doubt that "tomorrow" has arrived. In ecological research, it is now time to right the imbalance, to match the now-achieved more differentiated, more dynamic model of the environment with an equally differentiated and dynamic conception of the developing person. Perhaps in 10 years time Alan Pence will again exercise his discerning, integrative talents to create a sequel to this volume. The work will document an even more gratifying next stage in the life course of this research endeavor, one that reflects a matured scientific conception that does equal justice to the complexities of the person and of the environment, and, what is even more to be longed for, a stage that achieves the reality of a more humane environment for richly developing human beings.

REFERENCES

Bronfenbrenner, U. (1951). Toward an integrated theory of personality. In R. R. Blake and G. V. Ramsey (Eds.), *Perception: An approach to personality* (pp. 206–257). New York: Ronald Press.

Bronfenbrenner, U. (1970). *Two worlds of childhood: U.S. and U.S.S.R.* New York: Russell Sage.

Bronfenbrenner, U. (1974). Developmental research, public policy, and the ecology of childhood. *Child Development, 45,* 1–5.

Bronfenbrenner, U. (1975). Reality and research in the ecology of human development. *Proceedings of the American Philosophical Society, 119,* 439–469.

Bronfenbrenner, U. (1977a). Toward an experimental ecology of human development. *American Psychologist, 32,* 513–531.

Bronfenbrenner, U. (1975b). Lewinian space and ecological substance. *Journal of Social Issues, 33,* 199–213.

Bronfenbrenner, U. (1979). *The ecology of human development: Experiments by nature and design.* Cambridge, MA.: Harvard University Press.

Bronfenbrenner, U. (1986a). Recent advances in research on human development. In R. K. Silbereisen, K. Eyferth, & G. Rudinger (Eds.), *Development as action in context: Problem behavior and normal youth development*
(pp. 287–309). Heidelberg and New York: Springer-Verlag.

Bronfenbrenner, U. (1986b). Ecology of the family as a context for human
development. *Developmental Psychology*, *22*, 723–742.

Bronfenbrenner, U. (1986c). The war on poverty: Won or lost? *Division of
Child, Youth, and Family Services Newsletter*, *9*(3), 2–3.

Bronfenbrenner, U. (in press). Interacting systems in human development:
Research paradigms, present and future. In N. Bolger, A. Caspi, G.
Downey, & M. Moorehouse (Eds.), *Persons in context: Developmental
processes*. New York: Cambridge University Press.

Bronfenbrenner, U. and A. Crouter (1983). The evolution of environmental
models in developmental research. In P. H. Mussen (Ed., W. Kessen,
volume editor), *Handbook of child psychology: Vol. I. History, theory, and
methods* (pp. 357–414). New York: Wiley.

Thomas, W. I. and Thomas D. S. (1927). *The unadjusted girl*. Boston: Little,
Brown.

INTRODUCTION

ALAN R. PENCE

The terms "ecological" and "socio-ecological" have become increasingly common in research studies on child development and family functioning. In part, this increase is attributable to the work of Urie Bronfenbrenner, which culminated in his 1979 publication of *The Ecology of Human Development*. Bronfenbrenner noted that his work built on earlier traditions of ecological thought arising from several disciplines. Of particular importance to Bronfenbrenner was the work of Kurt Lewin and his development of the field theory in psychology. Other ecological pioneers identified by Bronfenbrenner included: George Herbert Mead and William and Dorothy Thomas in sociology; anthropologists Ralph Linton and Ruth Benedict; and, to a certain degree, John Dewey in education. Each of these individuals, and others from various disciplinary pursuits, shared a strong sensitivity to the importance and impact of the broader social ecology on human behavior and development.

Bronfenbrenner's purpose in promoting the ecological perspective throughout the 1970s (1974, 1977, 1979) was to shape a revitalized ecological approach into a significant alternative "force" in research on human development and its application in social programs and policy. In issuing his now classic condemnation of the "science of the strange behaviour of children in strange situations with strange adults for the briefest possible periods of time" (1979, p. 19), Bronfenbrenner hoped to tug loose from their moorings the firm lines connecting developmental research to highly controlled, lab-based studies.

The effect of Bronfenbrenner's criticism, within developmental research, led not to a complete throwing off of the traditional lines in favor of a new approach, but rather to a sufficient loosening of those established ties to allow an expanded ecological sensitivity. It is arguable that this loosening of the traditional "mooring lines" is all that was sought or attainable in Bronfenbrenner's call for a new approach to an understanding of human development; despite the presence of numerous definitions and propositions in *The Ecology of Human Development*, the thrust of the work is far more conceptual than methodological, more a call to thoughtful, systemic awareness than an operationalized, methodological guide. The diversity of examples cited by Bronfenbrenner in his text reflect and reinforce the interdisciplinary origins of ecological thought as well as underline that a range of methodological approaches are possible in undertaking ecologically sensitive research.

Some (Glossop, this volume) see in Bronfenbrenner's theoretical, conceptual, and methodological eclecticism the equivalent of "letting a hundred (intellectual) flowers bloom." These flowers of ecological thought took root in a wide variety of fields throughout North America, both in the social sciences and in the social services. The origin of this volume lies in the development of one research response to Bronfenbrenner's call to examine the social systems, and interactions of social systems, as they influence child development. The Victoria Day Care Research Project (Pence & Goelman, 1982) sought to better understand the impact of the *interaction* between caregiving and family microsystems on children's development. As the Victoria Project developed, the question of how other researchers in North America had understood the call for ecologically sensitive research and sought to operationalize their understandings became, itself, a focus of interest. Twelve individuals, representing a diversity of backgrounds and research involvements within the social services and social sciences, were invited to spend three days in Victoria, British Columbia, sharing their own experiences of the ecological road from concepts to methodology.

The diversity of backgrounds represented by the participants produced a range of orientations towards ecological thought. Some of the participants were very aware of Bronfenbrenner's work and could identify specifically how it had influenced their own thinking. Others had a limited awareness of Bronfenbrenner and traced their own orientation toward ecological thought through very different historical and disciplinary channels. In spite of the diversity of backgrounds and orientations, it became increasingly apparent that the models of ecological thought that had guided the individual inquiries and brought them together for discussion also provided a structure that enabled the participants to communicate ideas and experiences across the varied disciplines and backgrounds.

Central to communication among participants was an awareness and acceptance of a nested-systems model. That model, which places the microsystems of immediate experience within a mesosystem of two or more microsystems, is in turn embedded in the exosystem of non-immediate social contact; each of these system levels is, in turn, nested within a macrosystem of socio-cultural mores, values, and laws. Participants brought to the meeting, in the form of precirculated papers, studies that addressed one or more elements of ecological thought and design. The diversity of these studies initially appeared problematic—truly our "flowers" were not of one species. However, when one began to view the studies from the perspective of a nested systems model, not only could each be described using terms common to the model, but its relationship to other studies, similarly juxtaposed against the model, became evident.

To explain the workshop from a different perspective, the nested systems model provided the participants with a socio-ecological "map" possessing sufficient range and flexibility such that each participant's work could be identified as intersecting with one or more of the systemic rings of the map or addressing issues that were critical at one or more of the map's system levels. With such a relational perspective of one project or paper to another, the connections among the various participants could more easily be appreciated.

A related dynamic of the ecological model is that it is enhanced rather than diminished by difference and diversity. Unlike linear models, which largely depend on the degree of overlap among contiguous parts to make relationships comprehensible, the inherent holism of an ecological model provides its own relational structure (in this case, the nested systems), which is enriched and expanded through the diversity of specific studies embedded in that structure. As an example, studies that focus primarily on the interactions within and between the microsystems of nonparental care and the family are enhanced by studies examining the microsystems of family and workplace. These two mesosystem studies (care and family, and family and workplace), when considered in combination, each generate an exosystem component with potential relevance for the other.

The essays collected for this volume represent a range of child and family issues and a variety of approaches to ecological research; each occupies a unique place in the ecological web, yet each is connected in a number of ways to the others. They share a common awareness of behavior and development as interactive elements in a fluid and changing interplay among near and more distant social systems. They share a respect for the complexities of undertaking research in natural environments and a desire to create more socially meaningful descriptions of behavior. The essays generate questions of social policy and the efficacy of various social program interventions.

Given the holistic model of nested systems, the chapters could, theoretically, have been arranged in a number of sequences, for ultimately it is the whole that defines relationships in a systems perspective, not the "sausage-string" of contiguous connections. However, insofar as the editor has not found a way to bring all of the chapters to the reader's attention at once (so that the interplay of the whole is immediately apparent), a sequence has been established. The sequence should, however, be seen only to fulfill the role of an usher—that is, to bring the participants efficiently into this theater-in-the-round so that the interplay of the whole can begin.

OVERVIEW OF CHAPTERS

The first chapter, by Glossop, sets the stage for the text. The question of how the ecological theater is different from other arenas of thought and inquiry is uppermost in Glossop's commentary—in what ways is it unique and in what way is it tied to other traditions, to other ways of understanding? Glossop provides a historical and philosophical context for the other chapters.

The next two chapters, Chapter 2 by Goelman and Chapter 3 by Anglin, highlight different aspects of Glossop's discussion of characteristics of ecological research. Goelman's presentation of data from Pence and Goelman's Victoria Day Care Study highlights the developmental effects of interaction between systems, while Anglin addresses the key concept of personal meaning that an event may have for an individual and the relationship of that personal meaning to data collection in ecological research.

In Chapter 4 Johnson and Abramovitch examine the question of microsystem interactions in their exploration of parental job loss and its impact on personal and family variables. Lero, in Chapter 5, addresses a question that is central to both the Johnson and Abramovitch work and to a planned national survey by Lero, Pence, Goelman, and Brockman, namely, the ability of a survey-based study to capture truly ecological data. In considering the potential of national surveyors, Lero discusses seven major national surveys undertaken in the United States and Canada.

Lamb, in a study on teen-age parenting (Chapter 6) returns the reader to a consideration of social system levels in the presentation of data in an ecologically sensitive fashion. Through a juxtaposition of broad-scope, national data as background and more individualized microsystem data from a Utah study, the figure-ground theme of ecological thought is explored.

Both Powell in Chapter 7 and Cochran in Chapter 8 consider social program interventions from an ecological perspective. Powell's community-based, parent education and support program takes issue with the monolithic model of community-based intervention, that is, intervention that imposes a uniform set of expectations and experiences on diverse and heterogeneous populations. The Family Matters Project, developed in 1976, is one of the earliest, large-scale project interventions to be conceived specifically within Bronfenbrenner's ecological model. Cochran's analysis of the model and the project presents a compelling account of the rich potential in ecological interventions and of the problems associated with such a complex undertaking.

Whereas neighborhoods, families, and individuals form the principal foci of interest for the preceding eight chapters, the final two chapters enhance those perspectives by considering the development of a planned community, in Chapter 9 by Shera, and national social policy, by Wharf in Chapter 10. Shera's work represents one of the rare opportunities to evaluate social planning in a newly developed community. Using ecological and other models of planning and evaluation, Shera considers the early development of a planned community. Wharf rounds out the collected chapters and studies with an ecological consideration of national social policy. The unification of micro- and macro-level elements in our society is perhaps nowhere more evident than in social policy. Social policy decisions at the national level shape the experiences of individuals and groups at all system levels.

The chapters in this book represent a broad range of ecological thought and research approaches. The inspiration and conceptual antecedents are different for each one, yet there are characteristics in each that identify them as part of the ecological family. Those characteristics will emerge throughout the book and be more fully considered in the conclusion. The chapters are a manifestation of what James Garbarino described as a principal virtue of an ecology of human development, "its potential for a substantive eclecticism" (1984, p. 13). That potential is being increasingly realized throughout child and family research in North America. It is an eclecticism that represents more than a collection of token examples held in a tentative and fragile juxtaposition. It is rather a substantive eclecticism whose models require a diversity of perspectives in order to more fully reflect the complex interplay of personal and systemic influences that shape human behavior and development.

REFERENCES

Bronfenbrenner, U. (1974). Developmental research, public policy, and the ecology of childhood. *Child Development, 45,* 1–5.

Bronfenbrenner, U. (1977). Toward an experimental ecology of human development. *American Psychologist, 32,* 513–31.

Bronfenbrenner, U. (1979). *The ecology of human development: Experiments by nature and design.* Cambridge, MA: Harvard University Press.

Garbarino, J. (1982). *Children and families in the social environment.* Hawthorne, N.Y.: Aldine Publishing Co.

Garbarino, J. (1984). The evolution of an ecological perspective on human development. In J. Anglin (Ed.), *Report on a research workshop—Education and support for parenting: An ecological perspective on primary prevention.* Workshop Report. Ottawa: Health and Welfare Canada.

Lewin, K. (1951). *Field theory in social sciences: Selected theoretical papers.* New York: Harper & Row.

Pence, A. & Goelman, H. (1982). *Day care in Canada: Developing an ecological perspective.* Research proposal submitted to the Social Sciences and Humanities Research Council of Canada.

ECOLOGICAL RESEARCH
WITH CHILDREN AND FAMILIES

From Concepts to Methodology

1

Bronfenbrenner's Ecology of Human Development: A Reappreciation

ROBERT G. GLOSSOP

Robert Glossop's opening chapter sets the historical and philosophical context out of which Bronfenbrenner's definition of the ecological framework emerged. Glossop's consideration of research traditions in Western thought provides a rich introduction to the roots and present expressions of ecological research with children and families.

In 1979 Urie Bronfenbrenner's book entitled *The Ecology of Human Development: Experiments by Nature and Design* was published. While it would be premature to designate this work as a landmark, it is evident that the "new theoretical perspective" (p. 3) that Bronfenbrenner articulated then has already spawned work of lasting significance. For Bronfenbrenner, what was needed was a perspective that would permit the observation of interconnections between the processes of human development, the environments in which development occurs, and the reciprocal relations between the multiple environments within which any person develops.

An Ecological Framework For Theoretical Integration

Those who were prompted to pursue the "promise" of ecological research outlined by Bronfenbrenner have been inspired by his critique of the limitations of traditional approaches to the study of human development as encountered in the literatures of cognitive and developmental psychology. They have been inspired more by his redefinition of a field of inquiry and his conceptual orientation than by the specific theoretical and methodological directives he incorporated in the "theoretical schema" (Bronfenbrenner, 1979, p. 3) he constructed. Bronfenbrenner's primary goal is the articulation of an analytic basis for the theoretical integration of extant knowledge and prospective research about interrelated "structures and processes in both the immediate and more remote environment as it shapes the course of human development throughout the life span" (p. 11). His contribution to this "evolving scientific perspective" (p. 21), which he labels "the ecology of human development" (p. 21), takes the form of 14 definitions, 9 propositions and subpropositions, and 50 hypotheses. It is suggested that the proposed hypotheses will function as heuristic tools to "identify questions, domains, and possibilities believed worthy of exploration" (p. 15).

According to a protégé, friend, and colleague of Bronfenbrenner (Garbarino, 1984), *The Ecology of Human Development* is:

> Despite its formal style—which uses propositions and sub-propositions, and hypothesis—it is an intensely personal statement by a man who blends scholarship, social activism, scientific research, teaching, and showmanship into a compelling "fourth force" for social scientists and social service practitioners (with the first three forces being behavioristic, psychodynamic, and humanistic perspectives). (p. 6)

For Garbarino (1984), the ecology of human development was not, in 1979, a "systematic theoretical conception" (p. 6) and "it did not, and still does not, aspire to the status of a *theory* as we use the term in speaking of Freud's psychodynamic theory, Piaget's cognitive theory, or Skinner's reinforcement theory" (p. 6). The ecology of human development, at least according to Garbarino's (1984) familiarity with it and its principal proponent," does not attempt to provide a kind of substantive, or content description of development. Rather, it began as a critique of conventional developmental psychology; a critique first of content, and then increasingly of method" (pp. 6–7).

In commenting on the evolution of Bronfenbrenner's project since 1979, Garbarino (1984) continues: "Of late, it has become an effort to define a field of inquiry, and its principal use to date has been as a framework for organizing knowledge . . . , generating research questions . . . , and evaluating social policy" (p. 7).

It is not particularly surprising that it was Bronfenbrenner's point of view rather than his propositions, axioms, and hypotheses that were most instrumental in stimulating an energetic and committed group of researchers, policy analysts, and service providers to critically assess the ways in which the conceptual and disciplinary commitments characteristic of their areas of expertise have unnecessarily limited the horizons of their knowledge and their action. As Garbarino (1984) suggests, "the experimental ecology of human development is not really either a discipline or a substantive theory" (p. 13). The actual content of an ecology of human development is derived from a variety of disciplines, each of which is more specifically focussed on one of the conceptually demarcated systems—organism, microsystem, mesosystem, exosystem, macrosystem—around which Bronfenbrenner constructs his theoretical model of the ecology of human development. For Bronfenbrenner (1979): "The ecology of human development lies at a point of convergence among the disciplines of the biological, psychological, and social sciences as they bear on the evolution of the individual in society" (p. 13).

The ecological framework of analysis and interpretation, what others have termed the socio-ecological map, reveals the ways in which one disciplinary or conceptual approach takes as problematic precisely what is taken for granted by others, and vice versa. It becomes possible, then, to examine the relative contributions each makes to knowledge not only about its designated subject matter, but as well about the relationships between these domains of inquiry: the individual, the patterns of immediate sociation, and the systems of social interaction and institutionalization. Furthermore, it opens the possibility of identifying the contributions that studies that are not explicitly ecological can make to the development of an ecologically sensitive appreciation of human development and interaction.

True to the integrative intention that motivates Bronfenbrenner, the central virtue of his point of view is its focus on the reciprocal relationships between systems rather than on the properties and processes characteristic of any one system. Thus, in a systematic way, Bronfenbrenner strives to provide a foundation on which a genuinely interdisciplinary dialogue might be based, a dialogue through which what is known of biological and psychological processes might actually resonate with what is known of social-psychological processes, the dynamics of economic or political change, and the trends of social and cultural change.

Those who came to claim an allegiance to the ecological vision of reality and of research shared a common perception of the limitations of apparently "non-ecological" research as reductionist, fragmented, and overly specialized. They did, however, find it more difficult to define, in positive terms, what they shared in common in terms of conceptual underpinnings, theoretical axioms, and methodological preferences. The quick invocation of the apparent virtues of theoretical, conceptual, and methodological eclecticism, which followed soon after Bronfenbrenner's book, amounted to the notion of "letting a hundred (intellectual) flowers bloom." Although such an eclecticism was intuitively appealing, it must be remembered that it was precisely the fragmentation of competing bodies of knowledge that claimed to explain or contribute to the explanation of the processes of human development that had stimulated Bronfenbrenner to pursue his integrative project of an ecology of human development.

It may indeed be time to give up the search for a theoretical fountain of certainty (Aron, 1970, p. 332; Holman & Burr, 1980, p. 734; Rabinow & Sullivan, 1979, pp. 1–4; Taylor, 1979, p. 69; Vanier Institute, 1981, pp. 100–05). Yet, to do so forces one to confront many of the unexamined assumptions and customary beliefs in the character and virtues of the scientific enterprise that have, thus far, permitted researchers to continue in their independent labors, secure in the belief that knowledge is ultimately cumulative and aggregative, and leads progressively to firmer and sounder approximations of the properties and processes of the real world.

If one simply disregards the goal of theoretical integration with which Bronfenbrenner initiated his project and toward which he tried to specify a theoretical model of the relationship between systems and if one simply accepts the inevitability of eclecticism, one is left with the dilemma that characterizes the present state of our knowledge with its various frameworks, competing images of families, incompatible interpretations of developmental processes, discordant predictions, prognoses, and prescriptions, all of which are or can be based on credible and often highly sophisticated research (Tufte & Myerhoff, 1979, p. 2). And, such fragmented knowledge hardly serves as a firm foundation for our actions, interventions, policies, and programs. Although there seems to be no alternative to an eclectic approach, few can be sanguine about simply leaving the decisions as to which conclusions, which theoretical constructs, which predictions, and which policy recommendations will hold sway in the world of action to such factors as prevailing ideological sentiments, dogma, influence, prestige, and so on.

CONTEXT AND THE ECOLOGICAL FRAME OF REFERENCE

What, then, are the characteristics common to the research projects and analyses that claim to be compatible with an ecological point of view?

According to Bronfenbrenner (1979),

> The ecology of human development involves the scientific study of the progressive, mutual accommodation between an active, growing human being and the changing properties of the immediate settings in which the developing person lives, as this process is affected by relations between these settings, and by the larger contexts in which the settings are embedded. (p. 21)

It is Bronfenbrenner's emphasis on the "immediate settings" and the "larger contexts" in which the immediate settings and the developing person are embedded that is almost universally acknowledged as the cornerstone of the ecological frame of reference. Bronfenbrenner (1979) specifies for purposes of theoretical development this "extended conception of the environment" (p. 22), which is "considerably broader and more differentiated than that found in psychology in general and in developmental psychology in particular" (p. 22), as the "ecological environment" (p. 22). The ecological environment is "conceived topologically as a nested arrangement of concentric structures, each contained within the next" (p. 22). Other advocates of the ecological perspective simply refer to an emphasis on context as the hallmark of their ecological sensitivities.

It would seem, then, that attention to the context or the environment within which any of the processes of human development under investigation is situated is a necessary dimension of sound ecological research. Nevertheless, such an emphasis on context hardly serves to differentiate between an ecological study and other researches equally attentive to the determinant relationships or elective affinities between the behaviors or properties of conceptually demarcated units of analysis and the external forces, factors, causes, or variables that are conceived to explain those behaviors or properties. Indeed, there are many frames of reference not explicitly designated as ecological that have as their primary objective the observation, description, and explanation of precisely such relationships. Such contextually sensitive frames of reference, in fact, abound in the social sciences, albeit perhaps more evidently in those disciplines that focus directly on the processes of sociation, such as so-cial psychology, sociology, anthropology, linguistics, ethology, and the like. It seems, then, that an emphasis on context may be a nec-essary characteristic of ecological research but is not sufficient, in it-self, to identify how an ecological frame of reference is different from, say, the frameworks of structural-functionalism, environmentalism, systems theory, structuralism, and so on. On what basis, for in-stance, would one choose to specify one's research as ecological or socio-ecological as distinct from simply sociological or social-psy-chological research.

The notion of context serves a necessary though not sufficient role as a defining characteristic of ecological research; it is a major piece of the larger puzzle that is the ecological frame of reference. Still, what part does this notion of context play in the ecological frame of reference? In recognizing that any concept assumes its sig-nificance and meaning not through a process of formal definition, but rather in its relation to those other concepts that make up the framework within which it is utilized, one must ask what meaning does the idea of context assume within the larger ecological frame of reference.

Bronfenbrenner (1979) acknowledges that, on the surface, an ecology of human development resembles "social psychology on the one hand and sociology or anthropology on the other" (p. 12). He goes on, however, to differentiate the perspective he is articulating by emphasizing the "focus of the present undertaking on the phe-nomenon of *development-in-context*" (p. 13) and he suggests that the ecological approach is distinguished by its emphasis on the way the accommodation of a human organism to its immediate environment is influenced by "forces emanating from more remote regions in the larger physical and social milieu" (p. 13).

For Bronfenbrenner, then, what is crucial to an ecological per-spective as distinct from other contextually sensitive approaches is the *primary* place accorded to the interactions between systems. He distinguishes this from a primary emphasis on a designated phe-nomenon seen in relation to external forces that supposedly influ-ence its properties and processes of development or change. As he states, "in ecological research, the principal main effects are likely to be interactions" (Bronfenbrenner, 1979, p. 38). If we speak of ecological factors as distinct from simply social or systemic factors, it is because our analytic interest has shifted from the epistemolog-ically improbable task of describing the essential and invariant properties of discrete phenomena as "things in themselves," uncontaminated by the influence of external factors, toward an appreciation of how the properties we attribute to any phenomenon are a function of perceived relations between it and its context. Thus, "in ecological research the investigator seeks to 'control in' as many theoretically relevant ecological contrasts as possible within the constraints of practical feasibility and rigorous experimental design" (Bronfenbrenner, 1979, p. 38).

The notion of context, then, replaces the more conventional notion of a particular stimulus acting as a singular cause to produce a predictable outcome or change in the state or properties of a discrete phenomenon. Such a proposition departs radically from our analytic traditions, which have been predicated on the notion of the aggregation of constituent parts that are defined without reference to the context of their relationships with other parts. Within that classical tradition, knowledge is thought to advance through the development of ever more sophisticated research techniques designed to shield the hypothetical part and its essential nature from the interference or contamination of its relations with other parts of its environment, its context. Thus, Bronfenbrenner, while not discarding the virtues of rigorous experimental design and indeed of laboratory settings in pursuit of *some* questions, strongly advocates more emphasis on naturalistic methods of observation and research. After all, the notion of experiment is etymologically related to the term "experience," yet it has come to convey an image of a context-free laboratory setting in which the circumstances within which experience takes place are artificial, to say the least. Analysis is predicated on a belief in the autonomous existence of discrete parts that hypothetically combine together in a simple additive fashion to form a whole, that is, the whole equals the sum of the parts.

THE PRIMACY OF ECOLOGICAL RELATIONS

Yet today, even physics, from which the social sciences have, rightly or wrongly, borrowed this mechanistic conception of reality (and of the knowledge that seeks to describe it), has abandoned the notions of causes and of discrete, isolatable, independent essential "things" or phenomena. According to David Bohm (1981), "analysis into autonomously existent elements is not relevant" (p. 133). The kind of description that is required by modern physics is, Bohm suggests, based on a new conception of order, a conception not predicated on the presupposition that there are fundamental and discrete essential essences that combine to constitute matter. Instead, attention is now directed toward the pattern, the context, as primary and responsible for the characteristics displayed by any analytically distinguished "part." As he says, "In so far as what is relevant *is* the pattern, it has no meaning to say that different parts of such a pattern are separate objects in interaction" (Bohm, 1981, pp. 133–34).

It is, for example, the circuit that determines whether a "part" of that circuit serves as a switch, or, to use another analogy, it is the relationship between apparently discrete sounds that determines whether a sound is a phoneme or noise. For the quantum physicist, "flow is, in some sense, prior to that of the 'things' that can be seen to form and dissolve in this flow" (Bohm, 1981, p. 11). Thus, he continues, the "atom" is nothing more then a "poorly defined cloud, dependent for its particular form on the whole environment, including the observing instrument" (p. 9).

This assertion of the ontological primacy of relations is radical and fundamental. It carries with it the potential to challenge our entire understanding of knowledge and the methods appropriate to its acquisition. It requires of us descriptive capacities that illuminate patterns and not parts. It tells us that it is the context that determines the content. It tells us that we require a new notion of order as contained in, encompassed within, and implicit in each so-called part or region of space and time. It tells us that notions of causation and predictability are regarded as useful for the examination of only some limited dimensions of our experience; there is an order that is not, in essence, related to predictability. It tells us that "to be is to be related." It tells us that the whole is not simply the sum of its parts.

Essential to the ecological perspective is its emphasis on context. To the extent that context is almost infinitely variable, this assertion of the primacy of contextual relations effectively defies the traditional scientific objective of answering, with certainty and on the basis of generalizable law-like statements, questions about the particular behaviors of discrete phenomena. Confined more to the realm of probability, one is forced to agree with Garbarino (1984) when he suggests that:

> The infuriating thing about all this (and the source of its creative and analytic power) is that almost *everything* in the context of development is variable, almost nothing is fixed, and the answer to most questions of the sort "Does X affect Y" is "It depends!" (p. 10)

Although radical and challenging in its epistemological and methodological implications, even the notion of the primacy of relations is not sufficient to distinguish an ecological frame of reference from other established conceptual frames of reference. Structuralism, in particular, shares with an ecological frame of reference this fundamental principle expressed by James Ogilvy (1981) in the following way: "Rather than beginning with simple *things* and then spinning networks of relations as secondary appendages, structuralism takes relations as constitutive of their *relata*" (p. 267). Similarly, the phenomenological tradition of philosophy challenges the pervasive influence that conventional analytic distinctions between parts and wholes have exercised on social thought in general and social research in particular. In the foreword to her translation of Martin Heidegger's *Identity and Difference*, Joan Stambaugh (1969) wrote:

> What is new about this understanding of identity as a relation is that the relation first determines the manner of being of what is to be related and the how of this relation. It is perhaps difficult for us to think of a relation as being more original than what is related, but this is what Heidegger requires of us. (p. 12)

THE PHENOMENOLOGICAL VALIDITY OF ECOLOGICAL RESEARCH

It is no accident that Bronfenbrenner's notion of context coincides with the phenomenological philosophy of Heidegger and the latter's appreciation of relations as primary. Indeed, for Bronfenbrenner (1979), the validity of ecological research hinges on the extent to which a research study proceeds explicitly from a phenomenological sensitivity to "the subject's definition of the situation" and to the "knowledge and initiative of the persons under study" (p. 32). In formal terms, Bronfenbrenner (1979) defines ecological validity as "the extent to which the environment experienced by the subjects in a scientific investigation has the properties it is supposed or assumed to have by the investigator" (p. 29).

Time and again, throughout the early pages of *The Ecology of Human Development*, Bronfenbrenner emphasizes that the "phenomenological conception of the environment . . . lies at the foundation of the [ecological] theory" (p.23). In doing so, he acknowledges his debt to Edmund Husserl, Wolfgang Köhler and

David Katz in, respectively, phenomenology and psychology, George Herbert Mead and William Isaac Thomas in sociology, Harry Stack Sullivan in psychiatry, John Dewey in education, and Ralph Linton and Ruth Benedict in anthropology. Most particularly, he builds on the phenomenological concepts of the "life space" and the "psychological field" as incorporated into the systematic theory of Kurt Lewin. Building on this tradition, Bronfenbrenner consistently stresses that ecological research is primarily interested in understanding how the processes of human development are influenced by the interrelations of the developing person and the multiple contexts in which he or she lives *as these contexts are experienced by that person*. Thus, Bronfenbrenner (1979) implores his reader to acknowledge that "the scientifically relevant features of any environment include not only its objective properties but also the way in which these properties are perceived by the persons in that environment" (p. 22).

Even the very definition of human development provided by Bronfenbrenner is thoroughly phenomenological, focussing as it does on the primacy of human perception, the conscious construction by the developing person of a subjectively meaningful image of the environment, and the efficacy of one's perceptions and conceptions as a basis for human agency relative to the intentions, goals, and meanings that motivate action and are embedded in observable behaviors. Bronfenbrenner (1979) acknowledges that it is the conception of human development as incorporated within his ecological theory that is "most unorthodox" (p. 9). He writes: "Here the emphasis is not on the traditional psychological processes of perception, motivation, thinking, and learning, but on their *content—what* is perceived, desired, feared, thought about, or acquired as knowledge" (p. 9). And, he argues that the most important elements in the environment that influence human development are those that "have meaning to the person" (p. 22) and that what is important to understand is the environment "as it is *perceived* rather than as it may exist in 'objective' reality" (p. 4).

Although he concludes that there are inherent limits to the researcher's capacity to obtain a "complete picture of the research situation as perceived by the participants" (p. 33) and that "ecological validity is a goal to be pursued, approached, but never achieved" (p. 33), Bronfenbrenner (1979) insists that the establishment of some degree of correspondence between the subject's and the investigator's view of the research situation—a correspondence he labels "phenomenological validity" (p. 33)—is "not only desirable but essential . . . in every scientific inquiry about human behavior and development" (p. 30).

It is essential because, for Bronfenbrenner, the processes of human development are the processes through which human life becomes meaningful and through which the environments in which life is lived become meaningful—that is, full of human meanings. When the focus of inquiry and research shifts from the formal processes of perception, motivation, thinking, and learning toward the substantive content of *"what* is perceived, desired, feared, thought about, or acquired as knowledge" (Bronfenbrenner, 1979, p. 9), the possibility of understanding human actions by subsuming them under context-free laws of social behavior in keeping with the goal of universally valid predictive theory à la the Newtonian cosmology is precluded. To incorporate the category of human meaning, in a meaningful way, into our researches requires an appreciation of and attention to the particular circumstances out of which the human attribution of meaning emerges. This necessity to acknowledge the particular socially, historically, and culturally relative expressions of meaning departs from the predominant conception of scientific knowledge as progressing "by abstracting from the particulars of experience only what is capable of an increasingly more universal explanation and by abandoning the rest to the domain of the contingent and the inscrutable" (Unger, 1975, p. 201).

The paradox that inheres in the Western tradition of analytic thought has led to a conception of science (along with corresponding methodological directives) that legitimates only those forms of knowledge that develop by becoming steadily more abstract, general, and formal. Knowledge is thought to progress toward the elegance of a purely formal mathematical system of propositions at the expense of failing to illuminate the concrete particularity, the rich substance, and the texture of experience. The world as portrayed through the generalizing and universalizing categories of our most "advanced" science is a world devoid of content, substance, and meaning. At the level of substance or concrete experience, our analytic traditions emphasize separation, division, and difference, while at the level of supposedly explanatory theory, it is sameness and generality that are emphasized without any hope of returning to experiential or particular references. For Unger (1975), the prevailing conception of the scientific enterprise "parades before the mind a vast apparatus of forms in which the particularity of experience is wiped out" (p. 201). Or, as John O'Neill (1975) succinctly expresses the same thought, "the scientist has no 'here'" (p. 44).

But, of course, the subjects of our researches do have a "here" and a "now." Indeed, their perceptions and their conceptions of the environment are contingent on the immediate characteristics of the world experienced as meaningful by them in relation to their particular and very concrete hopes, fears, and goals. What Alfred Schutz (1973) expressed in terms of his "postulate of adequacy" (p. 44), Bronfenbrenner (1979) restates as the conditions to be met by any study that claims for itself ecological validity:

> To disregard the meaning of the situation to the research subject is to risk invalid conclusions both for research and, particularly in the study of human development, for public policy. To close one's eyes to this possibility is, therefore, to be scientifically and socially irresponsible. (p. 31)

CONCLUSION

Bronfenbrenner's commitment to the phenomenological foundation of his theory of human development is not merely arbitrary. It is a commitment that recognizes explicitly the extent to which prevailing conceptions of scientific explanation deny the variability of human circumstances and the relativity of human perceptions and conceptions of those circumstances. By recognizing, as did Max Weber (1949), that the development of science, like the development of a person, is situated within a constellation of value-relevant concerns and public policy issues, Bronfenbrenner insists that it is the phenomenological sensitivity to the subject's perceptions and conceptions of the environment that provides the only bridge over the schism between descriptive and explanatory psychology that has, according to Michael Cole (1979), plagued the discipline since the time of Wundt and Dilthey.

We have, as Cole (1979) suggests, chosen not to follow Dilthey's "very enticing view of adequate psychological description" (p. viii) for the good reason that "the infinite tangles of past experience and present circumstances that make us what we are smother us in particulars, defying explanation or generalization" (p. viii). Nevertheless, having been guided more by Wundt's modest aspirations for psychology, Cole (1979) acknowledges that explanatory psychology has lost its ground, its raison d'être, its relevance: "The limitations of theory imposed by that choice do not rest easy. We are faced with the paradox of a successful science that tells us precious little about the concerns that beckon us to it" (p. viii).

It is in response to this dilemma—a dilemma that Edmund Husserl (1970) labelled "The Crisis of European Sciences"—that Bronfenbrenner proposes a theoretical framework that promises to reintegrate abstract bodies of knowledge, each of which informs us of only a partial and limited dimension of human experience and human development. To this end, he provides a formal theory complete with a conceptual language, theoretical propositions, and empirically verifiable hypotheses. Furthermore, he provides a range of methodological directives and suggestions with regard to, for instance: the generalization of findings across settings; the relative strengths and weaknesses of laboratory and field research; the stratification of research samples according to the ecological domains of interrelated systems; the status of the laboratory environment as a basis for comparison with developmental processes occurring in natural environments; the techniques of naturalistic observation; the generation of hypotheses; participant observation; and the secondary interpretation of primary research, to mention only a few.

Bronfenbrenner's commentary on these and other methodological issues is consistently based on a post-Kantian rejection of objectivist correspondence theories of truth and method, according to which scientific knowledge is supposed to be an accurate representation of objective reality stated in analytic terms by reference to abstract and general laws. In their place, he has offered a phenomenologically based invitation to return to the particular and natural circumstances within which human development is situated and to an appreciation of how the processes of human development and the processes through which knowledge of human development is secured necessarily involve the collaborative engagement and mutual respect of subjects and researchers.

The ecology of human development is more than a new conceptual framework within which we can continue to do what comes naturally. The notions of context and environment as they take their place within Bronfenbrenner's framework go beyond what Arne Naess (1973) has called a "shallow environmentalism" by according ontological significance to relations and interactions. The phenomenological definition of human development and the corresponding definition of ecological validity invoke an epistemological tradition that redefines the relationship between researcher and subject, and their respective contributions to the creation of knowledge; the relationship between knowledge, and public policy and value-relevant issues and concerns; and the nature of objectivity in social scientific research and the methods appropriate to its attainment.

Together, it is these ontological and epistemological assumptions, which find their expression in Bronfenbrenner's project, that can reunite those bodies of knowledge that illuminate the general and universal dimensions of human experience with those that inform us of its unique attributes. Bronfenbrenner's phenomenological commitments acknowledge the necessary unity of the particular and the universal. His attention to the multi-layered circumstances within which human development takes place, his acknowledgment of the human initiative and creativity involved in the interpretation of the meaning of those circumstances, and his criticism of the artificiality of research settings all serve as recognition that there is "no circumstance in which the universal can be abstracted from its particular form. It always exists in a concrete way" (Unger, 1975, p. 143). As such, Bronfenbrenner joins William Blake (1932, p. 1068) in challenging us to escape "From Single Vision & Newton's Sleep!" and instead,

> To see a World in a Grain of Sand
> And a Heaven in a Wild Flower,
> Hold Infinity in the palm of your hand
> And Eternity in an hour. (Blake, 1932, pp. 118–21)

It is now time, after the publication of *The Ecology of Human Development* in 1979, to reappreciate the ambitious aspirations with which Bronfenbrenner launched a project that genuinely promises to illuminate, in a rigorous manner, the "past experiences and present circumstances that make us what we are" (Cole, 1979, p. viii).

REFERENCES

Aron, R. (1970). *Main currents in sociological thought II: Durkheim, Pareto, Weber*. Trans. R. Howard & H. Weaver. Garden City, N.Y.: Doubleday.

Blake, W. (1932). *Poetry and Prose of William Blake*. G. Keynes (Ed.), New York: Random House.

Bohm, D. (1981). *Wholeness and the implicate order*. London: Routledge and Kegan Paul.

Bronfenbrenner, U. (1979). *The ecology of human development: Experiments by nature and design*. Cambridge, MA: Harvard University Press.

Cole, M. (1979). Foreword. In U. Bronfenbrenner, *The ecology of human development: Experiments by nature and design*. Cambridge, MA: Harvard University Press.

Garbarino, J. (1984). The evolution of an ecological perspective on human development. In J. Anglin (Ed.), *Report on a research workshop—Education and support for parenting: An ecological perspective on primary prevention* (pp. 3–18). Ottawa: Health and Welfare Canada.

Holman T. & Burr, W. (1980). Beyond the beyond: The growth of family theories in the 1970's. *Journal of Marriage and the Family*, 42(4), 729–37.

Husserl, E. (1970). *The crisis of European sciences and transcendental phenomenology: An introduction to phenomenological philosophy*. Trans. D. Carr. Evanston, IL: Northwestern University Press.

Naess, A. (1973). The shallow and the deep long range ecology movements. *Inquiry*, 16, 95–100.

Ogilvy, J. (1981). From command to co-evolution: Toward a new paradigm for human ecology. In. R. C. Schultz & J. D. Hughes (Eds.), *Ecological consciousness: Essays from earthday X colloquium*. Washington D.C.: University Press of America.

O'Neill, J. (1975). *Making sense together: An introduction to wild sociology*. London: Heineman.

Rabinow, P. & Sullivan, W. (1979). The interpretive turn: Emergence of an approach. In P. Rabinow & W. M. Sullivan (Eds.), *Interpretive social science: A reader*. Berkeley, CA: University of California Press.

Schutz, A. (1973). Common-sense and scientific interpretation of human action. In M. Natanson (Ed.), *Alfred Schutz: Collected papers, Vol. I: The problem of social reality*. The Hague: Martinus Nijhoff.

Stambaugh, J. (1969). Introduction. In Martin Heidegger, *Identity and difference*. New York: Harper & Row.

Taylor, C. (1979). Interpretation and the sciences of man. In P. Rabinow & W. M. Sullivan (Eds.), *Interpretive social science: A reader*. Berkeley, CA: University of California Press.

Tufte, V. & Myerhoff, B. (1979). Introduction. In V. Tufte & B. Myerhoff (Eds.), *Changing images of the family*. New Haven, CT: Yale University Press.

Unger, R. (1975). *Knowledge and politics*. New York: Free Press.

Vanier Institute of the Family. (1981). *A mosaic of family studies*. Ottawa: Vanier Institute.

Weber, M. (1949). Objectivity in social sciences and social policy. In M. Weber, *The methodology of the social sciences*. Trans. E. A. Shils & H. Finch. New York: Free Press.

2

The Relationship Between Structure and Process Variables in Home and Day Care Settings on Children's Language Development

HILLEL GOELMAN

A focus on interactions between systems is a distinguishing characteristic of most ecological research. Hillel Goelman, in his work with Alan Pence on the Victoria Day Care Research Project, examines the interaction of structure and process variables within the two microsystems of home and day care settings as those variables affect indices of children's language development.

Over the past 10 years researchers have increasingly acknowledged the complexity of studying the effects of day care on young children. Previous research approaches had conceptualized all nonparental care as a uniform set of experiences and effects for all children. These research efforts focussed largely on the question of whether or not "day care" was beneficial or detrimental to children's development. Subsequent research efforts, identified by Belsky (1984) as a second wave of day care research, began to compare the effects of different forms of day care. While these "second wave" studies appropriately focussed more attention on the diversity of day care programs and experiences in which children participate, Bronfenbrenner's (1977, 1979) ecological model of human development began to direct attention and inquiries to a wide range of individual, family, and societal factors that both contextualize and influence the effects that specific day care settings have on the children in those settings. In this chapter we explore a number of factors in children's home and day care settings that appear to contribute to aspects of the children's development. We also consider these findings in the broader contexts of societal and governmental factors that outline the types of day care services provided and the ways in which families select day care settings for their children.

While day care researchers have begun to address a wider range of mediating variables, many of the studies have focussed on these variables in isolation. Belsky (1984) has argued that relatively little research has addressed the three-way interaction among the social structure of the day care setting (type of care, number of children, adult : child ratio), process variables in the day care environment (daily experiences of the children), and the children's performance on developmental outcome measures. Goelman and Pence (1985, 1987a, b) developed this line of argument by suggesting the need for information on family structure variables (family membership, income, parental education levels) and family process variables (experiences, interactions) in the child's home environment (see Table 2.1). In a study of children, parents, and caregivers in three types of day care, Goelman and Pence (1987a,b) reported that children's poor performance on measures of expressive and receptive language was associated with certain family structure variables. There were also indications in those data that many of the children appeared to be entering day care at a later age than children from more advantaged home environments and, more troubling, were entering poorer quality day care settings. In this chapter we examine the relationships between a number of structure and process variables in the children's home and day care environments and the children's performance on standardized measures of expressive and receptive language development. We will consider the data from an ecological perspective in order to examine the interaction of microsystem, mesosystem, exosystem, and macrosystem factors that appear to contribute to aspects of the child's language development.

While a thorough review of the effects of family process and structure variables on child development is beyond the scope of this chapter, a number of significant studies in this area should be cited. Hetherington, Cox, and Cox (1979) found that family structure, in this case single-parent families in the period following a divorce, was associated with specific kinds of negative social behaviors in preschool aged boys and to a more limited degree in preschool aged girls. Bee, Barnard, Eyres, Gray, Hammond, Spietz, Snyder, and Clark (1982) found that children's IQ and language scores were correlated with levels of maternal education, stress, and perceived social support in her community. Studies of intact two-parent families

TABLE 2.1 Research on Structure and Process Variables in Home and Day Care Settings

	FAMILY					DAY CARE				HOME-DAY CARE CONTINUITY	VARIABLES CONTRIBUTING TO CHILD DEVELOPMENT
	STRUCTURE			PROCESS			STRUCTURE		PROCESS		
	1	2	3	4	5	6	7	8	9	10	
	SES	Maternal Education	1 Parent/ 2 Parent	Lang. & Soc. Interactions	Attitudes	Type Care*	Size/Ratio	Quality	Lang. & Soc. Interactions		(variable numbers)
Brofenbrenner et al 82	X		X		X						
Carew				X		HC-CDC			X		4, 6
Clarke-Stewart 81				X		HC-CDC-LFDC-UFDC			X	Soc. Int.	
Cochran 77				X		HC-FDC-CDC			X		4, 6
Cochran & Robinson 83			X	X		CDC-FDC-HC			X	Soc. Int.	3, 6
Corsaro 79				X		Nurs. School			X	Lang. Int.	
Cross et al 84				X		CDC			X		4, 9
Everson et al 84	X		X	X	X	HC-FDC-CDC			X		5
Fowler 78	X				X	CDC	X	X	X	Soc. Int.	7, 8
Goelman & Pence 87 a,b	X	X	X		X	CDC-LFDC-UFDC		X	X	Attitudes	1, 2, 3, 6
Goelman 86 a,b						FDC			X		
Golden et al	X	X		X		HC-FDC-CDC			X		4, 6
Heath 83		X		X		Elem. School					4
Hetherington et al 79			X			Nurs. School			X		3
Honig & Wittmer				X		CDC			X		
Howes et al 84				X		CDC		X	X	Soc. Int.	
Long 84				X		FDC-HC			X	Soc. Int.	
McCartney 82, 83, 84	X	X				CDC	X	X	X		1, 2, 8, 9
Pellegrini 84						Nurs. School			X		
Prescott et al 75						CDC	X	X	X		
Ruopp et al 79						CDC	X	X	X		7, 8, 9
Schwarz 83	X	X		X		CDC-FDC-HC	X	X			1, 2, 7
Stallings & Porter 80						LFDC-UFDC	X	X	X		
Stith & Davis 84	X	X		X	X	FDC-HC	X		X	Soc. Int.	
Tizard & Hughes 84	X			X		Nurs. School			X	Lang. Int.	
Wells 81	X			X		Primary Grades					4

* CDC = Center Day Care; FDC = Family Day Care; LFDC = Licensed Family Day Care; UFDC = Unlicensed Family Day Care; HC = Home Care

(Lamb, 1980; Smith & Daglish, 1977; Tauber, 1979; Weinraub & Frankel, 1977) have reported different kinds of play in mother–child and father–child interactions, with differences also apparently dependent on the gender of the child. Maternal employment patterns and maternal attitudes toward work have also come under study. Bronfenbrenner, Henderson, Alvarez, and Cochran (1982); Stuckey, McGhee, and Bell (1982); and Thompson, Lamb, and Estes (1982) have studied the relationship among maternal attitudes toward employment, employment situation, and the nature of mother–child interactions during periods of change in employment situations. These studies suggest the complexities involved in the interaction of structure and process variables in children's home environments, which serve both as context to and intervening variable with the effects of structure and process variables in children's day care environments.

While increasing numbers of researchers have directed their attention to structure and process variables in day care settings, very few studies have examined the three-way linkages among structure, process, and developmental outcomes. Rubenstein and Howes (1979) and Ruopp, Travers, Glantz, and Coelen (1979), for example, studied the relationships between day care structure variables and children's performance on developmental outcome measures. Goelman (1986a, 1986b); Prescott, Jones, and Kritchevsky (1967); Ruopp et al. (1979); and Stallings and Porter (1980) examined the relationships between day care structure and day care process. Carew (1980) and Golden, Rosenbluth, Grossi, Policare, Freeman, and Brownlee (1978) have studied a third side of the triangle, the relationship between the children's daily experiences in day care and their performance on developmental outcome measures. Studies conducted on children in home care, family day care, and center care in Chicago (Clarke-Stewart, 1981; Clarke-Stewart & Gruber, 1984) and Sweden (Cochran, 1977; Cochran & Robinson, 1983; Gunnarsson, 1978) examined the influence of a wide range of family and day care structure and process variables on children's developmental status. These studies included observations of children's social interactions at home and in day care, as well as the children's performance on standardized measures of social and intellectual development. Taken together, the results of these studies suggest that the confluence of such factors as maternal marital status, the gender of the child, the kinds of social interactions in which the child engages both at home and in care, and the type of day care setting in which the child is enrolled, contribute to the child's social and intellectual development. Among the strengths of both studies were the range of day care settings studied and the inclusion of naturalistic and structured observations as well as the administration of standardized measures of child development.

Research on the Bermuda Day Care Project (McCartney, 1983, 1984; McCartney, Scarr, Phillips, Grajek, & Schwarz, 1982; Schwarz, 1983) has addressed a number of relevant dimensions of the children's home and day care settings and their impact on child development. McCartney's work, which focussed on day care centers, indicated that certain family structure variables (maternal education level) and day care structure (quality) and process variables (functional content of teachers' utterances) were associated with children's performance on the *Peabody Picture Vocabulary Test* (PPVT) (Dunn, 1979). Schwarz examined the test scores of children who had been in home care, family day care, and center day care before and after their third birthday. In creating a composite "family advantage" score, Schwarz covaried out aspects of family structure (parental income and education levels) in order to study the effects of the different child care settings at different ages. Schwarz reported higher scores on developmental outcome measures in children enrolled in family day care prior to their third birthday and in center day care after their third birthday. Both McCartney's inclusion of a wide range of family and day care variables in her multiple regression analyses and Schwarz's creation of the family advantage covariate represented rare but valuable attempts to examine the relationships between family background and day care variables.

Goelman and Pence (1987a, b) examined a wide range of family background variables in a study of the children, parents, and caregivers in three types of day care. Complementing and extending the findings reported on the Bermuda studies, the results of a series of correlational and regression analyses indicated that a number of specific family structure variables (maternal education, occupation, income, and marital status) were associated with children's scores on measures of both expressive and receptive language development. Further, there were indications that children of single mothers with low levels of education, occupation, and income were enrolled in lower quality family day care settings operated by caregivers with less formal training in child care. This finding drew attention due to the fact that caregiver training was also found to be a significant predictor of performance on the PPVT and the *Expressive One-Word Picture Vocabulary Test* (EOWPVT).

The major thrust of this chapter is to examine in greater detail the nature of the day care environments in which the children were enrolled in order to, first, describe the nature of the environments and the daily experiences the children have in these settings and, second, examine the relationships between the day care environments and the children's performance on measures of development outcomes. Finally, these data will be considered in the context of the family background variables cited above.

DESIGN OF THE STUDY

The data reported below are drawn from the Victoria Day Care Research Project. (For a detailed report on the design and methodology of this study, please refer to Goelman & Pence, 1987a; Pence & Goelman, 1987.) The subject pool included 105 child-parent-caregiver triads in three commonly used community-based types of day care: licensed center care (CDC), licensed family day care (LFDC), and unlicensed family day care (UFDC). Approximately equal numbers of boys and girls from one- and two-parent families were included in the three types of care (see Table 2.2). Children included in the study were either only or first-born children, had been in their current day care setting for at least the previous six months, and were in their current day care arrangement for at least 30 hours per week. All of the mothers of the children were either working, studying, or actively looking for work on a full-time (greater than 30 hours per week) basis. Due to the fact that most of the day care centers in the study did not accept children under three years old, the children in the CDC group were older (average of 50.5 months) than the children in either the LFDC (38.8) or the UFDC (39.8) group. While no other socio-economic or family variables were used in selecting the sample, a series of analyses of variance revealed that no significant differences existed between the families in the CDC, LFDC, and UFDC groups on a wide range of such characteristics, including the ages at which the children began day care and most indices of parental education, income, and occupation.

The research components of the study included structured interviews with the children's mothers and caregivers (Pence & Goelman, 1987) as well as the observation and developmental outcomes that are the major focus of this paper. The interviews covered a wide range of topics. Parents were asked questions regarding economic and demographic information about the family, their day care preferences and search strategies when looking for care for their chil-

TABLE 2.2 Parent, Child, and Caregiver Participants in the Victoria Day Care Research Project, Broken Down by Type of Care, Family Structure, and Sex of the Target Child.

	Parent and Children					Caregivers
	One–Parent		Two–Parent		Total	
	Boys	Girls	Boys	Girls		
CDC	14	13	15	11	53	25
LFDC	4	7	7	9	27	24
UFDC	7	7	6	5	25	24
Total:	25	27	28	25	105	74

dren, their perceptions of the day care provider and the day care setting, and their satisfactions and dissatisfactions with their current day care arrangements. The caregiver interview addressed many of the same topics as the parent interview and included a number of items regarding the caregiver's employment and training history. As discussed below, much of the information gathered from the interviews was subsequently used as background, grouping, and contextual information for the results on the developmental outcome measures.

The Early Childhood Environment Rating Scale (ECERS) (Harms & Clifford, 1980) and *The Day Care Home Environment Rating Scale* (DCHERS) (Harms, Clifford, & Padan-Belkin, 1983) were used to assess the physical environment and programmatic aspects of the day care centers and family day care homes, respectively. The two scales, similar in structure and implementation, consist of approximately 35 discrete items clustered into sub-scale scores in such areas as "Learning Activities," "Furnishings and Displays for Children," "Language–Reasoning Activities," and "Social Development."

The second major tool used in the day care observations was the Child Observation Form (COF), an instrument based on time and event sampling of the children's interactions in the day care setting. This instrument was based on observational systems developed by Prescott et al. (1975) and Stallings and Porter (1980), which were used in the National Day Care Study (of day care centers) (Ruopp et al.,

1979) and the National Day Care Home Study (Fosburg, 1981), respectively. While those studies included a range of activities found in both kinds of day care settings, the play categories of greater interest to this study were those identified by other researchers as being associated with child development outcomes. For example, McCartney (1982, 1983, 1984), had reported that among children in day care centers, the frequency of interactions involving "information" (the explicit communication, or teaching, of information) was predictive of children's performance on the PPVT. A further question that the COF was designed to address was the relationships between other activities that might reasonably be thought to correlate with higher levels of language competence, such as reading and dramatic play, and those that might not, such as television viewing and routine personal care.

The COF was designed so that observers equipped with electronic timing devices and earphones would observe the child for 5-second "windows" and record the corresponding behavior in the subsequent 25 seconds. Observers who successfully completed a training program and who achieved an inter-rater reliability level of 85 percent accuracy (on both the COF and two rating scales) were hired to conduct the observations. Observers visited the day care settings on two different days within a 10-day period for a total of 6 hours in the day care setting. The COF was used for an hour on each day of observation during the free play periods, generating 240 "windows" of activity for each child. The rating scales were completed at the end of the two days of observation.

The two major outcome measures used in this study were the PPVT for receptive language development and the *Expressive One-Word Picture Vocabulary Test (EOWPVT)* (Gardner, 1979) for expressive language. The tests were administered to the children on three occasions at six-month intervals over an 18-month period. Two other nonstandardized outcome measures were administered and are not included in the data analysis reported in this chapter. A brief child interview was also conducted in order to give the child an opportunity to describe the day care experience from his/her own personal perspective.

In the section below we consider the results on three measures: the day care structure, as indicated on the rating scales of the physical environment and day care "program"; the day care process, as indicated by the children's interactions and daily experiences; and the children's performance on the first administration of the measures of expressive and receptive language.

RESULTS

As reported in detail in Goelman and Pence (1987b), analyses of variance (ANOVA) on the standardized PPVT indicated that the difference between the mean UFDC scores (93.4) and the mean CDC (101.2) and LFDC (101.1) scores approached significance as revealed by a main effect for type of care ($F = 2.29, p < .10$). Further, the main effect for family structure ($F = 3.54, p < .06$) suggested that the difference between the mean scores for children in one-parent families (96.3) and two-parent families (102.1) also approached significance. Similar patterns were revealed on the ANOVA on the EOWPVT, where the UFDC scores (98.6) were significantly lower ($F = 3.11, p < .04$) than those for children in the CDC (107.3) and LFDC (109.3) settings. The higher mean scores of the children from two-parent families (107.4) over the children from one-parent families (104) was not significant ($F = 0.73, p < .39$). In neither the PPVT nor the EOWPVT ANOVAS were significant interactions reported.

In order to pursue the findings that children in the UFDC group scored lower on both language measures than children in the other two groups, further analyses were undertaken to examine the quality of the day care environments and the nature of the activities, interactions, and experiences in which the children engaged. Analyses of variance on the activities in which the focal children engaged revealed interesting patterns of similarity and difference between and across the three types of care. Children were found to engage in certain "typical" early childhood activities in approximately equal proportions in the CDC, LFDC, and UFDC settings. Thus, no significant differences were found on the percentages of time spent on structured fine motor, art and music, dramatic play, reading, and gross motor activities in the three types of day care. On other activities, however, significant differences were found. Children in CDC, for example, engaged in significantly more "information" activities (9.5%) than did children in either LFDC (3.1%) or UFDC (5.1%, $F = 7.89, p < .001$). Conversely, children in UFDC engaged in watching educational television (7.6%) significantly more than did children in either LFDC (3.8%) or CDC (.01%, $F = 11.65, p < .001$). There were also significantly fewer episodes involving routine personal care in UFDC settings (10.51%) than in either LFDC (16.39%) or CDC environments (15.62% $F = 3.74, p < .02$).

As shown in Table 2.3, performance on the developmental outcome measures tended to correlate more with aspects of the family day care settings than with the day care center environments. The

TABLE 2.3 Correlation of *PPVT* AND *EOWPVT* Scores with Selected Day Care and Family Variables ($P = .05$)

	1	2	3	4	5	6	7	8	9	10	11	12	13	14	15	16
1) PPVT Scores	—															
2) EOWPVT Scores	.75	—														
3) FDC-Learning Activities		.33	—													
4) FDC-Social Development	.51	.48		—												
5) FDC-Language Development					—											
6) FDC-Total Quality	.33	.32	.79	.88	.72	—										
7) CDC-Language Activities							—									
8) CDC-Creative Activities								—								
9) CDC-Social Development									—							
10) CDC-Total Quality							.82	.86	.87	—						
11) Information Activities (CHOBS)			.36	.41	.46	.44	.33	.35	.28	.27	—					
12) Caregiver Education						.31				.31		—				
13) Maternal Education (LFDC)													—			
14) Maternal Education (UFDC)														—		
15) Maternal Education (CDC)	.37	.33													—	
16) Maternal Education (All 3)	.34	.31										.27				—

25

only CDC variable with which both the *PPVT* and *EOWPVT* correlated
was the level of maternal education among the mothers of children
in center day care. It is interesting to note that the frequency of "information" interactions was significantly correlated with specific sub-
scales as well as total quality ratings of both the family day care and
center day care settings. The differing patterns of correlations between the family day care and center day care groups were comple-
mented by a series of regression equations that examined the effects
of specific day care and family variables identified in Goelman and
Pence (1987a, b), that is, type of care and maternal level of education. Due to the different structure and scoring in the *ECERS* for the
CDC ($\overline{M}$ = 171.9, SD = 22.26) and the *DCHERS* for the LFDC ($\overline{M}$ = 114,
SD = 22.17) and the UFDC ($\overline{M}$ = 96.04, SD = 23.84) groups, the center care and family day care groups were examined separately. These
analyses revealed that for the children in family day care, the *DCHERS*
rating scores were significant ($p < .05$) predictors of 11 percent of the
variance of the children's *PPVT* scores and that for the children in
center care maternal education level was a significant ($p < .05$) predictor of 14.4 percent of the variance.

Taken together, these sets of analyses suggest that family background, type of day care, and differing patterns of interactions in the
day care settings are associated with differences in the test scores of
the children. Specifically, aspects of the quality of family day care
appear to be more closely associated with test score performance than
do the parallel aspects of quality in the day care centers. In order to
examine this pattern in greater detail, a set of qualitative data analyses were undertaken to complement the quantitative analyses reported above. A computer search was conducted to identify family
day care and day care center environments that, on the basis of their
total quality ratings, were either one standard deviation above the
mean or one standard deviation below the mean *DCHERS* and *ECERS*
scores. When the settings had been identified, a wide range of information on the children, the environments, and the activities
within the environments were produced. While these data are of a
purely descriptive nature and inferential tests of significant differences would be inappropriate due to the absence of any criteria of
randomness, the differences between the high and the low quality
FDC and CDC settings are quite instructive.

Of the 15 high quality family day care homes (mean quality score
= 134.9), 13 were licensed and 2 were not, while of the 11 low quality homes (mean quality score = 69.6), 9 were unlicensed and only
2 were licensed. Children in the high quality FDC settings had mean

scores of 98.4 and 109 on the *PPVT* and *EOWPVT*, while children in the low quality family day care homes has scores of 83.3 and 86.6, respectively. In terms of the types of activities in which they engaged, the children in the high quality settings had higher frequencies than children in the low quality settings in information (6.3 vs. 1.5), reading (7.7 vs. 4.9), structured fine motor (10.3 vs. 6.6), and gross motor (16.9 vs. 6.4) activities. Conversely, children in the low quality FDC settings watched more educational television (12.4) and noneducational television (5.4) than did children in the high quality FDC environments (3.0 and .99, respectively).

In sharp contrast, the high and low day care center environments were much more similar to each other than were the high and low family day care settings. While the mean rating score the higher CDC settings (206) far exceeded that of the lower CDC settings (143.2), both the *PPVT* scores of the higher (99.2) and lower (104.2) settings and the *EOWPVT* scores of the higher (197) and lower (107.2) settings are strikingly similar. Children in the higher CDC group were observed to engage in more information activities (10.7) than children in the lower settings (5.9), but children in the lower CDC settings engaged in more reading (13.0) than children in the higher CDC settings (6.0). Further, children in the lower quality CDC settings engaged in more structured fine motor (14.0 vs. 7.2), art and music (19.0 vs. 13.0), and dramatic play (26.7 vs. 19.1) activities than children in the higher quality settings. No instances of television watching, either educational or noneducational, were observed in either the higher or the lower CDC environments.

DISCUSSION AND CONCLUSIONS

The patterns of results revealed by the quantitative and qualitative analyses reported above strongly suggest that main effects or monocausal explanations for the children's performance on measures of expressive and receptive language development cannot accurately reflect the complex interaction of factors in the microsystems, mesosystems, and exosystems of day care that appear to have an impact on aspects of the children's development. While factors within each of these systemic levels contribute to developmental outcomes, the challenge that is presented within an ecological perspective is to examine the ways in which factors in the family and the day care setting together and separately affect the focal child.

The effects of family background reported above and in Goelman and Pence (1987b) are of particular note in light of the striking similarities of the families using the three types of day care. When the various family and socio-demographic variables were analyzed separately, there were similar distributions of one- and two-parent families using the three types of day care and of the equivalent mean levels of maternal education, occupation, and income, and paternal levels of occupation and income. Further, the children were all enrolled in care for approximately the same number of hours per week and had all been in their current care arrangements for approximately the same amount of time. Despite these similarities, the clusterings of certain family background characteristics suggested that low resource families, as characterized by the combination of single parenthood and lower levels of maternal education, occupation, and income, were associated with lower scores on their children's PPVT and EOWPVT results. In confirmation of much of the literature in this area, these data lend further support to the importance of family structure variables in a microsystem of major importance to the developing child.

More troubling, however, was the finding that disproportionate numbers of children from low resource families were also entering low resource day care environments. These day care settings, the other major microsystem within which the children participated, also appear to contribute differentially to the child's performance on standardized measures of child development. While the quality of family day care homes was found to be related significantly to the children's test scores, the quality of rating of the day care centers in this study was not. However, in both the center and family day care settings, significant correlations were found between indices of the quality of the day care environments (as suggested by the sub-scale and total scores of the ECERS and DCHERS) and the frequency with which the focal child participated in developmentally facilitative activities in the day care setting. When children from low resource families are entering low resource day care settings and the results point to detrimental effects on the children's development, a portrait of an unsupported ecology of family life emerges from the data.

To a large extent these findings replicate and complement those reported in McCartney's analysis of the home and day care center influences on children's language development (McCartney, 1983, 1984; McCartney et al., 1982). As in the current study, McCartney found that maternal education level was a significant predictor of children's performance on the PPVT, that the frequency of "informational" utterances by the day care teachers was positively correlated with the overall quality of the day care settings, and that significant correlations existed between most of the sub-scale and total quality ratings on the ECERS. While McCartney reported a direct relationship between the ECERS and PPVT scores, this was not found for the day care centers in the Victoria Day Care Research Project. This relationship was found, however, between the rating scale scores for the family day care homes and both the PPVT and EOWPVT scores.

The identification of structure and process characteristics in home and day care settings that are associated with children's language development raises questions regarding mesosystem dynamics, specifically the effects of (dis)continuity between the child's experiences at home and in care. While the clustering of certain variables (single parenthood; low levels of education, occupation, and income; enrollment in lower quality family day care settings) can begin to inform the study of continuity, much more conceptual and empirical work is needed in order to both describe and analyze the precise nature and effects of continuity between home and day care settings. While family structure variables can begin to define the outlines of family dynamics, more information is needed on the nature of the process variable, the experiences, the activities, and the interactions of the persons in both their home and day care settings. Fine grained studies of language interactions conducted by Clarke-Stewart, 1981; Cross et al., 1984; and Tizard and Hughes, 1984 have revealed fascinating, yet contradictory, descriptions of the complexity, semantic contingency, language functions, and conversational competence in the child–child and child–adult discourse in home and day care settings. While Cross and her colleagues found striking similarities in the language environments of children's home and day care settings, Tizard and Hughes reported that language in the home was far more frequent and more "cognitively demanding" than talk in preschool settings. The extent to which the patterns of similarity and difference in the language environments of these settings are either positive or negative influences on the child's language development remains a question of intense debate. In language, as well as other aspects of continuity between home and day care (beliefs, values, perceptions), the extent to which effects of (dis)continuity are incidental, negligible, additive, cumulative, or exponential is a major challenge for researchers investigating the interaction of home and day care characteristics on children's development.

The patterns of results revealed in this study point to a series of questions regarding the effects of specific exosystem factors on the families and children involved. Day care services, funding, and licensing in many jurisdictions, including the one in which this study was conducted, are often influenced by, if not administered in the context of, welfare services to the poor. The fact that many children from relatively disadvantaged home environments found their way into lower quality day care settings, resulting in a kind of "worst of all worlds" situation and, apparently, lower scores on standardized measures of development, raises questions regarding the broader contexts of the ecology of day care in specific socio-demographic settings. This would entail a consideration of such factors as day care service delivery models, the availability of preferred modes of care, the existence and use of day care information service agencies, the interface of parents' employment related needs and the families' day care related needs, and the costs of different types of day care as well as the availability of subsidy to assist in paying those costs. Programs and standards regarding the training of day care providers and the licensing of day care facilities would also enter into the equation in order to address questions of how families select their day care arrangements from the options available to them and what factors help to determine the quality of those settings.

As Moncrieff Cochran and other participants in the Victoria Symposium have pointed out, the data in this chapter accentuate the "good news–bad news" implications of the availability of both formal and informal day care alternatives. Formal day care services, largely in the form of licensed day care centers, are relatively few proportionate to the need in a given socio-demographic setting, and the licensing procedures that set minimum standards of quality generally result in a narrow range of quality. Geared primarily to families that need day care on a full-time basis during regular working hours, the formal system offers what can be argued is a more restricted range of services to a more restricted clientele. The informal sector is allowed to float more freely. Parents can engage caregivers to provide care on a more individualized basis, tailoring the day care arrangements to their own particular needs. Parents preferring a more home-like arrangement to formal group care may seek out and appreciate what is available in the informal sector. Within this "freely floating" system, however, another crucial variable is allowed to float, and that is the range of quality care. As indicated by the results of this study, the quality of care varies tremendously, and when low quality day care compounds low quality home care, the child is clearly in a very disadvantageous position.

The data reported above strongly suggest that aspects of family structure, day care structure, and day care process influence the development of the young child. This tripod of information needs to be strengthened by the addition of data regarding the nature of developmentally facilitative processes, interactions, and experiences within the child's home setting. Specifically, information on adult–child language interactions in a range of family backgrounds would greatly enhance the clarity of the findings revealed in this study. This information from one of the major microsystems within which the focal child participates is necessary, but not sufficient, to gain an understanding of the complex interaction of factors within and across the systemic levels of the ecology of day care. Toward this end, research on "the effects of day care" must continue to focus on both the discrete pieces of the day care puzzle as well as the complex ways in which those pieces fit together.

REFERENCES

Bee, H. L.; Barnard, K. E.; Eyres, S. J.; Gray, L. A.; Hammond, M. A.; Spietz, A. L.; Snyder, C.; & Clark, B. (1982). Prediction of IQ and language skill from perinatal status, child performance, family characteristics and mother–infant interaction. *Child Development, 53,* 1134–56.

Belsky, J. (1984). Two waves of day care research: Developmental effects and conditions of quality. In R. C. Ainslie (Ed.), *The child and the day care setting* (pp. 1–34). New York: Praeger.

Bronfenbrenner, U. (1977). Toward an experimental ecology of human development. *American Psychologist, 32,* 513–31.

Bronfenbrenner, U. (1979). *The ecology of human development: Experiments by nature and design.* Cambridge, MA: Harvard University Press.

Bronfenbrenner, U.; Henderson, C. R.; Alvarez, W. F.; & Cochran, M. (1982). *The relation of the mother's work status to parents' spontaneous descriptions of their children.* Ithaca, NY: Department of Human and Development and Family Studies, Cornell University.

Carew, J. V. (1980). Experience and development of intelligence in young children at home and in day care. Monographs of the Society for Research in Child Development, 45 (6–7, Serial No. 187).

Clarke-Stewart, K. A. (1981). Observation and experiment: Complementary strategies for studying day care and social development. In S. Kilmer (Ed.), *Advances in early education and day care,* Volume 2 (pp. 227–50). Greenwich, CT: JAI Press.

Clarke-Stewart, K. A. & Gruber, C. P. (1984). Day care forms and features. In R. C. Ainslie (Ed.), *The child and the day care setting* (pp. 35–62). New York: Praeger.

Cochran, M. M. (1977). A comparison of group day and family childrearing patterns. *Child Development, 48,* 702–07.

Cochran, M. M. & Robinson, J. (1983). Day care, family circumstances and sex differences in children. In S. Kilmer (Ed.), *Advances in early education and day care,* Volume 3 (pp. 47–67). Greenwich, CT: JAI Press.

Corsaro, W. (1979). Sociolinguistic patterns in adult–child interaction. In E. Ochs & B. B. Schieffelin (Eds.), *Developmental pragmatics* (pp. 373–90). New York: Academic Press.

Cross, T.; Parmenter, G.; Juchnowski, M.; & Johnson, G. (1984). Effects of day care experience on the formal and pragmatic development of young children. In C. L. Thew & C. E. Johnson, *Proceedings of the Second International Congress for the Study of Child Language,* Volume II. New York: University Press of America.

Dunn, L. M. (1979). *Peabody Picture Vocabulary Test—Revised.* Circle Pines, MN.: American Guidance Services.

Everson, M.; Sarnat, L.; & Ambron, S. (1984). Day care and early socialization: The role of maternal attitude. In R. C. Ainslie (Ed.), *The child and the day care setting* (pp. 63–97). New York: Praeger.

Fosburg, S. (1981). *Family day care in the United States: Summary of the national day care home study.* Washington, DC: U.S. Department of Health and Human Services.

Fowler, W. (1978). *Day care and its effects on early development.* Toronto: Ontario Institute for Studies in Education.

Gardner, M. F. (1979). *Expressive One-Word Picture Vocabulary Test.* Novato, CA: Academic Therapy Publications.

Goelman, H. (1986a). The facilitation of language development in family day care homes: A comparative study. *Early Child Development and Care,* 25(1/2), 161–74.

Goelman, H. (1986b). The language environments of family day care. In S. Kilmer (Ed.), *Advances in early education and day care,* Volume 4 (pp. 153–79). Greenwich, CT: JAI Press.

Goelman, H. & Pence, A. R. (1985). Toward the ecology of day care in Canada: A research agenda for the 1980s. *Canadian Journal of Education, 10*(4), 323–44.

Goelman, H. & Pence, A. R. (1987a). Some aspects of the relationships between family structure and child language development in three types of day care. In D. L. Peters & S. Kontos (Eds.), *Annual advances in applied developmental psychology, Volume II: Continuity and discontinuity of experience in child care* (pp. 129–46). Norwood, NJ: Ablex.

Goelman, H. & Pence, A. R. (1987b). Effects of child care, family and individual characteristics on children's language development. In D. Phillips (Ed.), *Quality in child care: What does the research tell us?.* NAEYC Monograph Series, Washington, DC: National Association for the Education of Young Children, 89–104.

Golden, M.; Rosenbluth, L.; Grossi, M.; Policare, H.; Freeman, H.; & Brownlee, E. (1978). *The New York infant day care study.* New York: Medical and Health Research Association of New York City.

Gunnarsson, L. (1978). *Children, day care and family care in Sweden: A follow-up*. Gothenburg, Sweden: University of Gothenburg.

Harms, T. & Clifford, R. (1980). *The early childhood environment rating scale*. New York: Teachers College Press.

Harms, T.; Clifford, R.; & Padan-Belkin, E. (1983). *The day care home environment rating scale*. Chapel Hill, NC: Homebased Day Care Training Project.

Heath, S. B. (1983). *Ways with words: Language, life and work in communities and classrooms*. New York: Cambridge University Press.

Hetherington, E. M.; Cox, M.; & Cox, R. (1979). Play and social interaction in children following divorce. *Journal of Social Issues, 35*(4), 26–49.

Howes, C.; Goldenberg, C.; Golub, J.; Lee, M.; & Olenick, M. (1984). Continuity in socialization in home and day care. Paper presented at the Annual Conference of the American Educational Research Association, New Orleans.

Lamb, M. E. (1980). The development of parent–infant attachments in the first two years of life. In F. A. Pederson (Ed.), *The father–infant relationship: Observational studies in a family setting*. New York: Praeger Special Studies.

Long, F. (1984). Discontinuity in the social experiences of children in family day care homes. Paper presented at the Annual Conference of the American Educational Research Association, New Orleans.

McCartney, K. (1983). Many roads to heaven: The issue of quality in day care intervention programs. Paper presented at the Biennial Meeting of the Society for Research in Child Development, Detroit.

McCartney, K. (1984). Effects of quality of day care environment on children's language development. *Developmental Psychology, 20*(2), 244–60.

McCartney, K.; Scarr, S.; Phillips, D.; Grajek, S.; & Schwarz, J. C. (1982). Environmental differences among day care centers and their effects on children's development. In E. Zigler & E. Gordon (Eds.), *Day care: Scientific and social policy issues* (pp. 126–51). Boston: Auburn House.

Pellegrini, A. D. (1984). The effects of classroom ecology on preschoolers' functional uses of language. In A. D. Pellegrini & T. D. Yawkey (Eds.), *The development of oral and written language in social contexts* (pp. 129–44). Norwood, NJ: Ablex.

Pence, A. R. & Goelman, H. (1987). Silent partners: Parents of children in three types of day care. *Early Childhood Research Quarterly, 2*(2), 103–118.

Pence, A. R. & Goelman, H. (1987). Who cares for the child in day care? An examination of caregivers from three types of care. *Early Childhood Research Quarterly, 2*, 315–334.

Prescott, E.; Jones, E.; & Kritchevsky, S. (1967). *Group day care as a child-rearing environment: An ecological approach*. Pasadena, CA: Pacific Oaks College.

Prescott, E.; Jones, E.; Kritchevsky, S.; Milich, C.; & Haselhoef, E. (1975). *Assessments of child-rearing environments. Part I: Who thrives in group day care? Part II: An environmental inventory*. Pasadena, CA: Pacific Oaks College.

Rubenstein, J. L. & Howes, C. (1979). Caregiving and infant behavior in day care and homes. *Developmental Psychology, 15*, 1–24.

Ruopp, R. R.; Travers, J.; Glantz, F.; & Coelen, C. (1979). *Children at the center: Final report of the national day care study.* Cambridge, MA: Abt Books.

Schwarz, J. C. (1983). Infant day care: Effects at 2, 4 and 8 years. Paper presented at the Biennial Meeting of the Society for Research in Child Development, Detroit.

Smith, P. K. & Daglish, L. (1977). Sex differences in parent and infant behaviour in the home. *Child Development, 48*, 1250–54.

Stallings, J. & Porter, A. (1980). *Observation component of the National Day Care Home Study: Volume 3.* Washington, D.C.: U.S. Department of Health and Human Services.

Stith, S. M. & Davis, A. J. (1984). Employed mothers and family day care: A comparative analysis of infant care. *Child Development, 55*, 1340–48.

Stuckey, M. F.; McGhee, P. E.; & Bell, N. J. (1982). Parent–child interaction: The influence of maternal employment. *Developmental Psychology, 18*(4), 635–44.

Tauber, M. (1979). Sex-differences in parent–child interaction style during a free-play session. *Child Development, 50*, 981–88.

Thompson, R. A.; Lamb, M. E.; & Estes, D. (1982). Stability of infant–mother attachment and its relationship to changing life circumstances in an unselected middle-class sample. *Child Development, 53*, 144–48.

Tizard, B. & Hughes, M. (1984). Young children learning: talking and thinking at home and at school. London: Fontana.

Weinraub, M. & Frankel, J. (1977). Sex differences in parent–infant interaction during free play, departure, and separation. *Child Development, 48*, 1240–49.

Wells, G. (1981). *Learning through interaction: The study of language development.* Cambridge: Cambridge University Press.

3

THE PARENT NETWORKS PROJECT: TOWARD A COLLABORATIVE METHODOLOGY OF ECOLOGICAL RESEARCH

JAMES P. ANGLIN

In developing the Parent Networks Project, James Anglin encountered in a very personal way the issue of individual understanding that a subject brings to a research project. A major question raised in Bronfenbrenner's work, The Ecology of Human Development, *is "how the research situation was perceived and interpreted by the subjects of the study" (p. 30). That concern is the focus of Anglin's chapter.*

The Parent Networks Project (PNP) developed out of a social-ecological perspective on parenting and parent support. Unexpectedly, as the PNP process unfolded the author was led to abandon the original research methodology in favor of an alternative, more collaborative form of inquiry for which few guidelines were readily available. This chapter will focus on aspects of the PNP that appear to have significance for researchers attempting to understand and work within a social-ecological perspective on human development, regardless of the particular content focus.

In the field of psychology, the acknowledged father of the social-ecological perspective on human development is Urie Bronfenbrenner (Garbarino, 1982). In Bronfenbrenner's words:

> The ecology of human development involves the scientific study of the progressive, mutual accommodation between an active, growing human being and the changing properties of the immediate setting in which the developing person lives, as this process is affected by relations between these settings, and by the larger contexts in which the settings are embedded. (1979, p. 21)

The aspect of this formulation that appears to have captured the imagination of contemporary researchers to the greatest degree is the novel conception of the human environment. Bronfenbrenner points the way to settings, and interconnections between settings, that have traditionally lain outside the ken of the human development researcher. In Bronfenbrenner's now familiar terminology, the microsystem (the interrelations within the immediate setting), the mesosystem (the interrelations between microsystems), the exosystem (settings in which the person does not participate), and the macrosystem (patterns of culture and societal ideology) form "a set of nested structures, each inside the next, like a set of Russian dolls" (1979, p. 3). Using another analogy, Garbarino (1982) suggests that the ecological perspective on human development can be seen as offering "a kind of map for steering a course of study and intervention" (p. 25). The impact of this conception of the environment on the course of contemporary developmental research, social policy, and social intervention has been significant.

A second feature of the social-ecological formulation is its "unorthodox" (Bronfenbrenner, 1979) conception of development. "Development is defined as a person's evolving conception of the ecological environment, and his relation to it, as well as the person's growing capacity to discover, sustain, or alter its properties" (Bronfenbrenner, 1979, p. 9).

The emphasis that is put on the way the environment is perceived, and the meaning it has for the developing person, appears to have had much less impact on the research enterprise than it deserves. In Part One of *The Ecology of Human Development*, Bronfenbrenner makes important reference to this aspect no fewer than 20 times. Two quotations will illustrate the degree of importance attributed to this aspect by the author himself.

> Very few of the external influences significantly affecting human behavior and development can be described solely in terms of objective physical conditions and events; the aspects of the environment that are most powerful in shaping the course of psychological growth are overwhelmingly those that have meaning to the person in a given situation. (p. 22)

> This means that it becomes not only desirable but essential to take into account in every scientific inquiry about human behavior and development how the research situation was perceived and interpreted by the subjects of the study. (p. 30)

Clearly, Bronfenbrenner sees profound implications of his phenomenological perspective on human development for the carrying out of social science research. In the discussion that follows, the experience of the Parent Networks Project as it relates to issues of studying and understanding the perceptions of the subjects of research will be presented. As will become apparent, the learning that took place was of a highly personal nature, encompassing affective and moral as well as cognitive dimensions of the research enterprise.

BACKGROUND TO THE PARENT NETWORKS PROJECT

The PNP grew out of the author's review of the history, theoretical foundations, and current practice dimensions of parent and family life education in North America. As a result of this study, the author concluded that "the rich tradition of parent education suggests numerous opportunities for developing further the potential inherent in the parent education movement." Hope was raised that some breakthrough could be made in our understanding of, and response to, the needs of parents and families (Anglin, 1985). During the course of this review, the author became further convinced that any such breakthrough would likely entail the adoption of a social-ecological orientation to research and program development (see Figure 3.1). This conclusion was based largely on the state of the findings of evaluation research in the field. As stated by Harman and Brim (1980) in their survey of the field of parent education:

> In sum, available evaluation research remains inconclusive, although much less so than was the case over two decades ago. It can be stated emphatically that our understanding of parent education and our ability to plan its various components with greater effect are contingent upon, among other factors, the availability of much further research and investigation which should endeavour to uncover far more knowledge regarding its manifold, complex, and intertwined elements. (p. 258)

It appeared, then, that the parent education "mouse" was too big for the traditional research "mousetraps." In response to Harman and Brim's challenge, the author proposed an "eco-cube" model for evaluating parent education programs.

The eco-cube model attempted to offer a social-ecological framework capable of encompassing the complex array of intertwined elements characteristic of parent education efforts, as noted by Harman and Brim. It was evident that traditional evaluation studies had only explored a few, relatively gross dimensions of parent education efforts.

FIGURE 3.1   Ecological Framework for Evaluation of Parent Education Programs

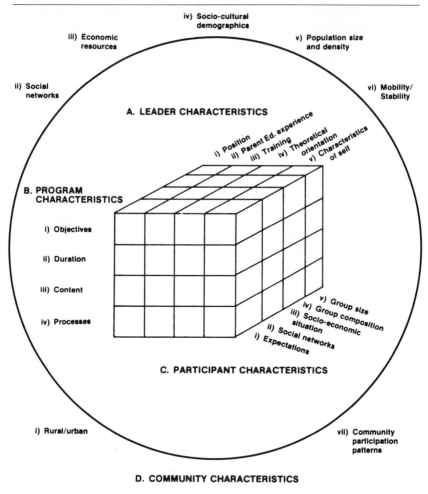

However, having attempted to design a blueprint for a better mousetrap, the author began to have serious doubts about this enterprise. After all, parent education had a 160-year history in North America. How could it be that all our research still added up to inconclusive findings? It seemed implausible that 160 years of history had simply been awaiting some variation of the eco-cube framework. These thoughts were further fueled by Kagan's (1984) review of the research in the child development field. He reported that "after a thorough examination of the evidence on family socialization, two respected psychologists [Maccoby and Martin] concluded that the relations between parental behavior and the child's qualities are generally ambiguous" (p. 275). As do Harman and Brim, Kagan concludes that a possible explanation is "that most of the research has not been sufficiently sophisticated" (p. 275). But he further suggests that

> Psychologists, in expecting to find a relation between what parents do and a particular outcome in the child, have generally failed to appreciate that the child is always interpreting the actions of parents. . . . Rarely will there be a fixed consequence of any single event—no matter how traumatic—or special set of family conditions. (Kagan, 1984, pp. 275–76)

This insight of Kagan's concerning the significance of the child's interpretation of the actions of parents seemed to echo Bronfenbrenner's admonition that researchers in the field of human development must "seek to discover empirically how situations are perceived by the people who participate in them" (1979, p. 24). It began to appear that the ambiguity and inconclusiveness of many previous research findings had more to do with the lack of sensitivity to the perceptions of the research subjects than with the number or interconnectedness of the various dimensions identified. The evolution of the PNP convinced the author that this was indeed the case. *More* research, as traditionally conceived, would never do the trick; a *different orientation* appeared to be called for.

THE PARENT NETWORKS PROJECT

One of the most exciting and promising areas of study uncovered in the author's review of the parent education literature was the burgeoning literature on social networks and social support. Much exploration was taking place with a view to enhancing the role of professional services in relation to family life (Bronfenbrenner & Cochran, 1976; Cochran & Woolever, 1983; Froland, Pancoast, Chapman, & Kimboko, 1981; Gottlieb, 1981, 1983; Powell, 1984; Whittaker & Garbarino, 1983). These wide-ranging studies demonstrated the need for professionals to better understand, and be more sensitive to, the informal helping networks existent in the community. As the goal of the literature review to this point had been to ascertain the most effective approaches to supporting parents and families that might be applicable in Victoria, B.C., the next step appeared to be the undertaking of a formal study to assess the informal and formal helping patterns of parents in the city. To this end, the services of two graduate students and a small grant for project resourcing were secured for a period of three months.

The intention was to use a modified version of a social network assessment questionnaire that had been developed by Kazak and Wilcox (1984) for use in a study of the social support networks of families with handicapped children. After several weeks of background preparation, team development meetings, and discussion of project assumptions and objectives, a draft questionnaire was prepared. The day came when I, as project coordinator, was to take the questionnaire home for proofreading prior to sending it for printing. I took the instrument with me, pencil in hand, as I slid into a warm bath to soak while I read. As I began, the notion came to *complete* the questionnaire as if I were a respondent, rather than simply to review it. This decision proved to be a decisive turning point in the project.

The page of demographic questions proved relatively unproblematic; however, the request for a list of the persons (initials and relationship to respondent) who were "an important source of help, support, or encouragement" proved to be a different matter. First, the question was followed by 28 lines down the page for the responses. As I began to write, those spaces overwhelmed me with their mute expectations; my tacit understanding told me I could not fill all those spaces. As I proceeded to complete the question, many insights into my world of informal support, some coming clearly to mind for the first time, and numerous realizations about the impact of the instrument on me—the respondent—came to light. Some of these are set out below.

1. My wife was clearly first on the list. (Would I be first on hers; would I be on it at all?!)
2. Who is number two? Hmmm. My children? No, a "real" respondent wouldn't likely put children on the list. Then a friend who lived 2500 miles away came to mind. I see this person once or twice a year, talk perhaps once a month on the telephone, and collaborate on articles on occasion. B.G. went on second. Not because of our usual exchanges, but because a little over six weeks before, we had suddenly grown very close—more like biological and spiritual brothers than friends. This had been a profoundly

moving experience for me, and one that promised to significantly alter much of my orientation to life. How do I list him? I decided on "B.G.—friend." That was probably how the researcher could best make sense of it. If I put "B.G.—friend/brother," the researcher would really be confused. While writing the response, I was acutely aware of how little the researcher analyzing my response could understand about this relationship and what it meant to me. It was technically correct, but humanly so devoid of real content and significance. And what if we had asked for "only those you see every 4 to 6 weeks," as some questionnaires do? Or for "only those who live within the radius of a one-hour drive," as others do? With those criteria, B.G. would not even have made it onto the list!

3. Then a pause. Who is number 3? It took a few minutes of reflection, then I hesitated after putting down the initials. He is a friend and a clergyman. I was hesitant to put "clergy," not feeling totally comfortable exposing my religious orientation to others I didn't know. As well, I wasn't sure which response I "should" put.

These were some of my first reactions. A full description of my reactions, in relation to the questionnaire process and instrument, as well as self-realizations, could fill several pages. In brief, it was very apparent that the data the questionnaire was gathering, and could be expected to gather, were pale and misleading representations of the reality the instrument was seeking to explore and they were, necessarily, seriously incomplete. Some way had to be found whereby the data provided by the respondents could be not only gathered, but adequately interpreted and understood. The degree to which the questionnaire completion process could bring forward disturbing emotional and psychological reactions in respondents needed to be addressed as well. The instrument was, despite appearances, far from benign in its impact. Following completion of the questionnaire, for example, I had a powerful dream about a key person who had not met the criterion as a supporter and who had thus been left off the list of supporters. This dream led to some important new understandings and altered behavior in relation to the person in question. It was not too farfetched to imagine the questionnaire completion process having an impact, possibly quite negative, on a respondent's relationship with a spouse, partner, or other relation.

As a result of this powerful experience of becoming a respondent to my own instrument, I urged the two co-investigators to complete the questionnaire as well. They, too, had similarly strong reactions. Subsequent discussions led us to have serious doubts concerning both the validity of such an instrument in relation to the complexity of the phenomena we were seeking to understand, and the ethics of sending it out in an anonymous fashion. Clearly, administering such a questionnaire meant intervening significantly in the lives of the respondents without being able to predict what kind of emotional or psychological impact this intervention might have.

In addition, we began to identify ourselves, as researcher-persons, as the most critical instruments in the project. Paralleling Bronfenbrenner's concern that researchers pay attention to the meaning of the situation as the research subject perceives it, we began to realize the degree to which our values, beliefs, judgments, preferences, and so on, would have an important impact on every aspect of the study—conceptualization, process, instrumentation, and analysis. We were at a standstill, seriously wondering if we would be able to find a way to approach parents that would be both valid and ethical in our own minds.

After another two weeks, involving some pilot testing with a group of staff in a local agency, we devised an approach that seemed to us to overcome our greatest concerns. First, we decided we would approach only parents who had already indicated a willingness to engage issues of personal social support by joining a support group of some kind. Second, a brief questionnaire would be used to focus attention and to serve as the starting point for a group discussion of the responses to the questionnaire. Third, we would offer to meet with individuals or groups following the discussion to deal with any aspect of the process or experience that they might wish to discuss. We believed the dialogue (approximately 1-1/2 hours) following the completion of the questionnaire (approximately 15 minutes) would allow an opportunity to explore the meaning of the participants' responses, thus providing a respondent-verified understanding of the data. By approaching only groups of parents and offering to engage in follow-up meetings either individually or as groups, we believed we could assist the participants to deal with any potential negative impacts of the research intervention.

Six groups involving a total of thirty-nine persons were involved in the PNP, with the group meetings ranging in size from three to ten respondents plus one or two investigators. The interviews ranged from one to two hours in length, with a mean of one-and-a-half hours. All discussion was audio taped and transcribed verbatim. At the time of this writing, analysis using several different approaches is being undertaken.

However, it is not the findings from the group interviews that are most relevant here, but rather the renewed understanding of the social-ecological research enterprise itself that emerged from the experience of undertaking the PNP.

TOWARD A COLLABORATIVE RESEARCH APPROACH

The PNP began with the presumption that a prestructured questionnaire could be administered to a sample (ideally random) of parents, the data analyzed (with no personal contact with the respondents), and valid conclusions drawn (with no verification by the respondents themselves). Along the way, ethical as well as methodological concerns were raised that suggested some significant limitations to the conventional model utilized for such inquiry. In essence, the conventional model of survey research appeared to treat the subjects as objects. The conventional approach kept person-to-person contact of researchers and subjects to a minimum, was cooperative only to the extent that respondents cooperated if *they* agreed to complete *our* questionnaire, and ignored the fact that administering the instrument was like hitting the respondents with a hammer without being able to know where we were hitting them, how hard we were hitting them, and with what personal effect.

A review of other recent studies on social support attested to how typical these characteristics were in studies involving human subjects. The emergent and collaborative nature of the research process that was developed in the PNP could now be viewed as a search for a more human form of human inquiry—an attempt to make social science more truly a social enterprise. In the words of Heron (1981):

> Persons in relation in their world symbolizing their experience of the value of the presented world constitutes a fundament of the human condition. Every science is just a special case of this symbolizing activity. When the subjects of a science are other persons, then the idea that the researcher's underlying value system can exclude, need not consult or consider or cooperate with the value system of the subjects, can only tend to generate alienated pseudo-truths about persons. For an authentic science of persons, true statements about persons rest on a value system explicitly shared by researchers and subjects, and on procedural research

norms explicitly agreed by researchers and subjects on the basis
of that value system. Hence . . . the model of cooperative inquiry.
(p. 33)

A number of dimensions of such a cooperative or collaborative
model of inquiry can be suggested on the basis of the PNP experience. These include:

1. The distinction between research and intervention becomes no
   longer tenable. Research involving persons directly is a significant form of intervention into their lives.
2. The separation between researcher and subject begins to dissolve. The PNP became a collaborative research enterprise with
   researcher-participants involved with participant-researchers.
   While roles and functions differentiated the two groups, they
   were free to exchange roles, either momentarily or for more extended periods of time. In the PNP dialogues, for example, the
   researchers were sometimes put in the position of being respondents to the questions of parents, while respondents freely
   proposed concepts and questions for exploration.
3. The human "instrument" comes to be seen as the most important instrument in the inquiry. In fact, the conduct of the inquiry
   sought to take advantage of the investigator–subject interactions
   rather than to minimize them. The "objective analyst" became the
   "engaged participant."
4. Dialogue becomes a central method in the research process. It was
   through person-to-person interactions that the meanings of
   words, concepts, and experiences were negotiated and mutually
   understood. One-way data gathering by means of a questionnaire proved inadequate to the task defined as critical by Bronfenbrenner, namely, to understand "the meaning of the situation
   to the research subject" (1979, p. 31).
5. Personal learning on the part of all inquiry participants (participant-researchers and researcher-participants) becomes an explicit goal. In the PNP, we came to realize, with Heron (1981), that:

   > Doing research on persons involves an important educational
   > commitment: to provide conditions under which subjects can en-
   > hance their capacity for self-determination in acquiring knowl-
   > edge about the human condition. (p. 35)

6. Qualitative review of concepts and themes replaces the quantitative analysis of questionnaire data as the prime source of understanding.

7. Ethical concern for the respondents moves beyond offering the option to choose not to complete or return the questionnaire to ensuring that appropriate supports are available to respondents if and when they feel the need for supports.

It is apparent that a number of tenets characteristic of conventional social scientific inquiry were contravened as the PNP evolved. These shifts, although unanticipated, were by no means accidental, but rather represent the outcome of adopting a phenomenological viewpoint, as advocated so forcefully by Bronfenbrenner. An explicit acknowledgment of the free, active, and purposive nature of the subjects, and all that this implies, led to the adoption of a collaborative stance and methodology. Of course, such a shift in emphasis is by no means unique. Such an approach has been characteristic of a number of streams within a variety of social science disciplines for some considerable time. Symbolic interactionism in sociology (Blumer, 1969; Mead, 1934), ethnomethodology (Garfinkel, 1967), gestalt in psychology (Kohler, 1929), and, of course, phenomenology in philosophy (Natanson, 1966; Schutz, 1966) represent but a few examples.

This rich, if alternate, tradition is currently being tapped by an emerging network of social scientists in North America and Great Britain (Lincoln & Guba, 1985; Reason & Rowan, 1981) who are questioning the adequacy and appropriateness of what they refer to as the "positivistic paradigm" for much of contemporary social science inquiry. The experience of the Parent Networks Project suggests that such an alternative form of inquiry may be well suited to a social-ecological orientation with its essential commitment to understanding the meaning of phenomena for the subjects of inquiry. Whereas the positivistic orientation is particularly well suited to the task of objective measurement, the "phenomenological" (variously referred to as "naturalistic," "ethnographic," and "postpositivistic") orientation provides an approach whose goal is subjective understanding.

Without attempting to undertake a thorough review of these two orientations, a number of fundamental ontological and epistemological axioms can be outlined as suggestive of their different world views. Lincoln and Guba (1985), under the headings of "positivistic paradigm" and "naturalistic paradigm," juxtapose five such axioms for consideration (see Figure 3.2).

While it is certainly simplistic to reduce social science research to two polar camps, the comparison can be helpful in bringing forward some eternal issues of "being" and "knowing" for consideration in the light of contemporary social-scientific issues and concerns.

FIGURE 3.2   Contrasting Positivist and Naturalist Axioms

Axioms About	Positivist Paradigm	Naturalist Paradigm
The nature of reality	Reality is single, tangible, and fragmentable.	Realitites are multiple, constructed, and holistic.
The relationship of knower to the known	Knower and known are independent, a dualism.	Knower and known are interactive, inseparable.
The possibility of generalization	Time- and context-free generalizations (nomothetic statements) are possible.	Only time and context-bound working hypotheses (idiographic statements) are possible.
The possibility of causal linkages	There are real causes, temporally precedent to or simultaneous with their effects.	All entities are in a state of mutual simultaneous shaping, so that it is impossible to distinguish causes from effects.
The role of values	Inquiry is value-free.	Inquiry is value-bound.

SOURCE: Lincoln and Guba, 1985, p. 37

CONCLUSION

The experience of the PNP suggests to the author that the axioms of the naturalistic paradigm deserve serious consideration by social-ecological researchers. In the social/behavioral sciences, one must ask where one can investigate a phenomenon about which one could assert a tangible reality; where one can assume independence of the observer, in any sense; where stability exists over all (over any!) time and context factors; where there is direct and unidirectional causality; and where there is freedom from value constructions.

For direct inquiry into the lives of persons, the naturalistic/phenomenological approach offers an alternative that can be carried out in a responsible and disciplined manner. Just as the terrain of how persons perceive the situations within which they find themselves is largely unexplored, so too are the methodologies for undertaking this exploration. It would be most fitting for social-ecological researchers to be in the forefront of this methodological task, seeking to complete the vision proposed by Bronfenbrenner in his influential statement of social-ecological theory.

REFERENCES

Anglin J. P. (1985). Parent education: Can an old tradition address new needs and new realities? In R. Williams, H. Lingren, G. Rowe, S. V. Zandt, P. Lee, & N. Stinnett (Eds.), *Family strengths 6: Enhancement of interaction*. Lincoln, NB: Center for Family Strengths, University of Nebraska—Lincoln.

Blumer, H. (1969). *Symbolic interactionism: Perspective and method*. Englewood Cliffs, NJ: Prentice-Hall.

Bronfenbrenner, U. (1979). *The ecology of human development: Experiments by nature and design*. Cambridge, MA: Harvard University Press.

Bronfenbrenner, U. & Cochran, M. (1976). The ecology of human development: A research proposal to the National Institute of Education, Cornell University.

Carr, W. & Kemmis, S. (Eds.). (1983). *Becoming critical: Knowing through action research*. Victoria, Australia: Deakin University Press.

Cochran, M. (1985). The parental empowerment process: Building on family strengths. In J. Harris (Ed.), *Child psychology in action: Linking research and practice*. London: Croom Helm.

Cochran, M. & Henderson, C. R. Jr. (1985). Family matters: Evaluation of the parental empowerment program. The Comparative Ecology of Human Development Project, Cornell University.

Cochran, M. & Woolever, F. (1983). Beyond the deficit model: The empowerment of parents with information and informal supports. In I. Sigel & L. Laosa (Eds.), *Changing families*. New York: Plenum Press.

Froland, C.; Pancoast, D. L.; Chapman, N. J.; & Kimboko, P. J. (1981). *Helping networks and human services*. Beverly Hills, CA: Sage Publications.

Garbarino, J. (1982). *Children and families in the social environment*. Hawthorne, NY: Aldine Publishing Co.

Garfinkel, H. (1967). *Studies in ethnomethodology*. Englewood Cliffs, NJ: Prentice-Hall.

Gottlieb, B. H. (1981). *Social networks and social support*. Beverly Hills, CA: Sage Publications.

Gottlieb, B. H. (1983). *Social support strategies: Guidelines for mental health practice*. Beverly Hills, CA: Sage Publications.

Harman, D. & Brim, O. G. Jr. (1980). *Learning to be parents*. Beverly Hills, CA: Sage Publications.

Heron, J. (1981). Philosophical basis for a new paradigm. In P. Reason & J. Rowan (Eds.), *Human inquiry: A coursebook of new paradigm research*. New York: John Wiley.

Kagan, J. (1984). *The nature of the child*. New York: Basic Books.

Kazak, A. E. & Wilcox, B. L. (1984). The structure and function of social support networks in families with handicapped children. *American Journal of Community Psychology, 12*(6), 645–61.

Kohler, K. (1929). *Gestalt psychology*. New York: Liveright.

Kuhn, T. S. (1970). *The structure of scientific revolutions* (2nd ed.). Chicago: University of Chicago Press.

Lincoln, Y. S. & Guba, E. G. (1985). *Naturalistic inquiry*. Beverly Hills, CA: Sage Publications.

Mead, G. H. (1934). *Mind, self and society*. Chicago: University of Chicago Press.

Natanson, M. (Ed.) (1966). *Essays in phenomenology*. The Hague: Martinus Nijhoff.

Powell, D. R. (1984). Enhancing the effectiveness of parent education: An analysis of program assumptions. In L. G. Katz (Ed.), *Current topics in early childhood education*, Vol. V. Norwood, NJ: Ablex.

Reason, P. & Rowan, J. (Eds.) (1981). *Human inquiry: A sourcebook of new paradigm research*. New York: John Wiley.

Schutz, A. (1966). *Collected papers, Vol. III: Studies in phenomenological philosophy*. The Hague: Martinus Nijhoff.

Schwartz, P. & Ogilvy, J. (1979). *The emergent paradigm: Changing patterns of thought and belief*. Analytical Report 7, Values and Lifestyles Program. Menlo Park, CA: SRI International.

Whittaker, J. K. & Garbarino, J. (1983). *Social support networks: Informal helping in the human services*. New York: Aldine.

4

PATERNAL UNEMPLOYMENT
AND FAMILY LIFE

LAURA C. JOHNSON
RONA ABRAMOVITCH

Laura Johnson and Rona Abramovitch are concerned both with the personal meaning that a difficult life event (in this case a job loss) has for an individual and with the interaction effects of that event with the microsystem of family life. Although such mesosystem dynamics would seem to be of fundamental social importance, few studies of paternal job loss have adopted an ecological frame of inquiry.

From the 1960s to the 1980s major changes have been seen in North American family life and in the settings external to the family in which family members participate. The traditional family roles described by functionalist theorists (Parsons, 1955) have begun to break down as more mothers become wage earners. Fathers have been encouraged to share responsibility for child rearing and housework; some researchers have described this new male role as intrinsically rewarding (Levine, Pleck, & Lamb, 1983; Pleck, 1982) or as supportive of women's new economic roles (Barnett & Baruch, 1986). Nonetheless, time-budget research indicates that the attitudes and behaviors of most North American males have changed relatively little. Most men see their own economic role of provider as salient, and loss of that role through unemployment can represent a very major disruption in their lives (Binns & Mars, 1984; Hakim, 1982; Ray & McLoyd, 1986).

Current economic and social conditions in North America are creating a situation in which labor market factors external to the family are forcing large numbers of families to change traditional family roles. This chapter describes research investigating how such change in work roles may affect men and their families. While there is much speculation about the social costs of unemployment, there is little specific information on the ways in which paternal unemployment affects families. We do not know, for example, whether unemployed fathers assume roles of child care provider and homemaker. For those that do become actively involved in parenting, does that role provide a source of gratification? Alternatively, do the responsibilities of parenting constitute an added source of stress for unemployed fathers?

Clearly, child care is a new, unanticipated role for most men who are thrust into it through job loss. To what extent do they feel competent and prepared for the parenting role? How does the length of time they spend in the parenting role affect men's feelings toward parenting? Does increased time spent in the role increase their learning about parenting and consequently their satisfaction with the role, or does duration increase the stress? What are unemployed fathers' feelings about the usefulness of formal community support services? Are there particular services or programs that they feel would be of value to unemployed parents and their children?

A further set of unanswered questions deals with the partner's role in the family and the workplace. How does she react to her spouse's changed work status? Does her partner's job loss cause her to move into a new or more demanding work role? Does she experience any difficulty accepting his increased involvement in child rearing? Does the degree of her involvement in a work role affect her partner's support for her parenting role? What about the effect of paternal unemployment on the marital relationship? Does unemployment affect the risk of divorce or separation? Is there evidence of increased tension or discord? Finally, how do the children fare in families with unemployed fathers? Is there any indication of negative impact on their behavior or development?

This research examines the various ways in which paternal unemployment may affect the roles and relationships within the family unit. The results of the study will begin to provide new information on the ways in which family systems adapt to parental unemployment. In a more general sense, the results of this survey will represent an important investigation into the flexibility of roles and relationships within the family.

This study originates from a conceptual framework that views work and family as two distinct but highly interrelated domains. Marked changes within both domains and in the relationship between the two have occurred in recent years. Interrelationships occur both at the micro-level of social roles and at the structural level between family and workplace. The present study examines the work–family nexus at the level of individuals who are involved in both work roles and parenting roles. Specifically, it focusses on the way in which loss of the work role through unemployment may affect fathers' parenting role.

Following a review of existing literature on paternal unemployment and families, this chapter will describe a pilot, qualitative phase of this research, followed by the results from a larger survey of unemployed fathers.[1]

PATERNAL UNEMPLOYMENT AND FAMILIES: RESEARCH EVIDENCE

Psychological Effects of Unemployment on Fathers

While there is a body of research linking unemployment with mental health problems (Dooley & Catalano, 1980; Fagin & Little, 1984), there is little evidence that relates specifically to unemployed fathers. There is no single study that examines the psychological impact of the simultaneous changes in the work role and the family role that characterize paternal unemployment. Depression era studies of the 1930s generally found a broad range of negative effects of unemployment on the psychological functioning of men and their families. However, individual reactions varied widely and appeared to be affected by factors such as age and sex of the unemployed person, the length of unemployment, the degree of economic loss, and family cohesion. (Reviews of the 1930s findings appear in Eisenberg & Lazarsfeld, 1938; Jahoda, 1979; and Liem & Liem, 1978.)

Although the Depression studies suggest possible consequences of contemporary unemployment, there are problems with generalizing those findings to the present due to profound changes in the social context in which unemployment occurs. More recently, Brenner (1973) used complex statistical procedures to determine that increased unemployment was associated with increased first admissions to mental hospitals in New York State. This finding has been criticized (Catalano, Dooley, & Jackson, 1981) on the grounds that

---

[1]Portions of this chapter appeared originally in the report "Between Jobs": Paternal Unemployment and Family Life, published in 1986 by the Social Planning Council of Metropolitan Toronto. This material is reprinted with permission from the Council.

economic decline may simply uncover existing untreated disorders, rather than produce symptoms in healthy people. (See also Catalano, Dooley, & Jackson, 1985.)

Additional research in Britain and the United States tends to support the relationship between unemployment and poor mental health. While none deals specifically with effects on fathers, it is possible to derive from these studies some general indications of factors that increase the negative impact of unemployment on mental health, emotional well-being, and life satisfaction. These factors include:

> being middle-aged, especially when married with young dependents (Liem & Rayman, 1982; Warr, 1978);
> having lower occupational status, i.e., semi-skilled or unskilled work (Hepworth, 1980);
> viewing work as central to one's identity and wanting to work (Swinburne, 1981; Warr, 1978);
> being unemployed for longer periods of time (Hepworth, 1980);
> having no warning of or control over job loss (Swinburne, 1981);
> not having financial resources to fall back on (Powell & Driscoll, 1978).

In a recent review of literature on the consequences of unemployment, Hakim (1982) observes that the psychological impact of job loss through unemployment is generally greater for men than for women (see also Schwefel, John, Potthoff, & Hechler, 1984), and generally greater for married than for single men. For married men, she observes, the paid employment and/or occupational role is a central focus of personal identity.

A Canadian study by Burke (1985), which compared the consequences of job loss for men and women, found more severe economic consequences for women, such as shorter duration of subsequent employment, lower earnings, and greater drop in wages. As a group, the unemployed in this study reported no more psychosomatic symptoms than the general population; however, they reported less satisfaction with life and less marital and family satisfaction. They also smoked more and drank more alcohol than the general population. German research conducted by Schwefel, John, Potthoff, and Hechler (1984) found poorer mental but not physical health among the unemployed in comparison with employed controls five to six weeks after becoming unemployed. The authors concluded that unemployment may remove the worker from an unhealthy environment or may affect mental health sooner than physical health. Results of this study also indicate that both financial difficulties and social isolation increase the negative impact of unemployment on mental health.

One of the very few recent studies to concentrate on the impacts of paternal unemployment on fathers and their families is an ongoing American study by Margolis and Farran (1984, 1985). Preliminary results of their research indicate that for fathers there is an association between duration of unemployment and depression, low self-esteem, anxiety, and resentment.

Effects of Paternal Unemployment on Families

Both historical and contemporary studies have shown male unemployment to have a variety of disruptive effects on family life. American studies of the 1930s, largely based on individual family case studies, found that unemployment had negative consequences for family relationships and functioning (Eisenberg & Lazarsfeld, 1938; Moen, 1983). More recently, Hakim (1982) reports that unemployment of a male breadwinner undermines family stability, increases friction between spouses, and may increase the risk of divorce.

Time-series analysis by South (1985) has shown that there is a rise in the divorce rate during periods of unemployment (although the author does not interpret this finding as being due to disruptive effects of unemployment on family life).

In a study of working class family life of unemployed men in Glasgow, Scotland, Binns and Mars (1984) report that the resulting drop in social and economic status combines with the upheaval in domestic roles to make the unemployed family a highly unstable unit.

Research by Thomas, McCabe, and Berry (1980) in the United States suggests that the negative consequences of paternal unemployment may be less severe among managerial and professional groups with somewhat older children. Asked, "How has your being unemployed affected your relationship with your spouse?" almost half (48%) of the middle-aged respondents reported no change, about one-third (37%) reported a negative effect, while 15 percent reported some positive effect. This finding suggests that, not surprisingly, a family's economic resources may cushion the impact of unemployment.

There is some evidence suggesting negative consequences for

the partners of unemployed males. A study done in the United States by Liem and Rayman (1982) found that negative consequences for wives of unemployed men increased with length of unemployment. After one month of unemployment, wives of the unemployed were somewhat more depressed than wives of the employed, but there were no other significant differences between the two groups. After four months of unemployment, wives of the unemployed were significantly more depressed, anxious, and sensitive about interpersonal relationships. Liem and Rayman (1984) subsequently report a variety of other family-level effects among families of unemployed males. Notable among these were shared perceptions by both spouses of increased family conflict and disorganization following unemployment, and increased risk of separation and divorce.

Preliminary results from another American study of unemployed men and their families (Margolis & Farran, 1984, 1985) also suggest that paternal job loss has adverse effects on the moods of both parents, with wives showing some evidence of increased depression and anxiety. In addition, British research by Fagin and Little (1984) found changes in the wives of unemployed men, such as increased depression and cigarette smoking. Interestingly, some wives benefitted from their husbands' unemployment by assuming a more important position in the family, derived from a new role as primary breadwinner.

One important contribution of the Fagin and Little study is the demonstration that the consequences of unemployment for the family depend on each family's developmental stage and particular mode of functioning. This research provides preliminary descriptive information about changes in family members' roles and relationships and the organization of family life that follow unemployment.

Generally, however, the literature on the impact of unemployment on the family contains little if any analysis of the way in which externally imposed changes in the work role may affect roles within the family. One exception is the recent study of male unemployment among blue-collar families in Glasgow, cited above (Binns & Mars, 1984), which does examine reciprocal effects of changes in highly traditional work and family roles. We need to address the same questions within the context of contemporary Canadian society.

Effects of Paternal Unemployment on Children

Children may also show the effects of paternal unemployment. The unemployed respondents in Liem and Rayman's 1982 research reported signs of stress in their children, including moodiness and problems in school and in peer relationships. British studies by Fagin (1981) and Fagin and Little (1984) reported similar results. It may be that children are susceptible only to the effects of prolonged parental unemployment. Early results from Margolis and Farran's ongoing longitudinal study of paternal unemployment (1984) found no significant health or behavioral consequences for children of unemployed fathers. The researchers speculate that children may be protected from immediate effects, while the significant effects may become manifest only in the longer term.

In summary, while there is growing consensus that unemployment is a major social problem with considerable social cost, there is relatively little research evidence on the direct effects of paternal unemployment on parenting and family life. The evidence that does exist suggests a number of areas of potential problems, some factors that may exacerbate the negative impacts of unemployment, as well as factors, such as financial security, that may cushion such effects. In fact, the conclusions from available research give considerable cause for concern about the social effects of paternal unemployment. The findings that married men experience particularly high levels of psychological distress from unemployment, that negative consequences for families seem to increase with the duration of fathers' unemployment, and that financial insecurity may increase the stress associated with unemployment all suggest that paternal unemployment may have a serious, negative impact on parenting and family life.

The present research was done in two phases. The first, more qualitative phase involved in-depth interviews with a small sample of unemployed fathers. Follow-up interviews, conducted six months after the initial interviews, examined the effects of duration of unemployment on parenting and family life. The second phase of the research involved surveying a large sample of unemployed fathers, using a structured research instrument developed from the initial, qualitative interviews. Results of these phases of the research are described below.

PHASE ONE:
A QUALITATIVE PILOT STUDY

The first phase in this research was a pilot study involving in-depth personal interviews with a sample of 30 currently or recently unemployed fathers of young children from birth through six years of age in an urban Canadian setting (Johnson & Abramovitch, 1985).

Respondents in this pilot study were located predominantly through "on-the-street" screening of men observed with young children in shopping malls, parks, and other public places.

The pilot study interviews included a large number of unstructured questions designed to elicit men's attitudes toward unemployment, toward child care and other family responsibilities, and toward support services. Analysis of the data from the pilot study led to the development of a series of structured interview items and scales designed to assess fathers' attitudes and reactions toward unemployment.

Results of that pilot study provided a preliminary portrait of the unemployed father, his needs, and his problems, and suggested a statistical relationship between the duration of paternal unemployment and certain aspects of the fathering role. Specifically, the pilot study suggested that for those fathers who assumed responsibility for child care, the longer the duration of the fathers' unemployment, the more likely they were to describe their children in negative terms.

Other findings from the pilot study were as follows:

• Job loss was found to be a major shock to the self-esteem of the fathers interviewed. Most of the fathers reacted to unemployment in negative terms, emphasizing its deleterious effects on themselves and other family members. Boredom, depression, and anxiety characterized many fathers' reactions to unemployment.

• A number of the unemployed fathers felt themselves to be stigmatized and blamed by the community at large. Loneliness and social isolation were reported by some of the fathers.

• On the positive side, one-half of the fathers interviewed in the pilot study cited opportunities to spend time with their children as one of the few positive consequences of their unemployment. On the negative side, some of the unemployed fathers found themselves suddenly thrust into child care and homemaker roles for which they felt ill-prepared.

• Many of the fathers noted a need for child care services to enable them to search for employment, as well as to have some time to themselves, away from their young children.

The Pilot Study Sample Six Months Later

The apparent association between duration of fathers' unemployment and their negative descriptions of their young children was one of the main findings of the pilot study. Since the duration of unemployment appeared to be such an important factor, a decision was made to re-interview the pilot sample six months after the original interview. Like the initial interview, the follow-up interview probed men's feelings about work and unemployment, their children, and their family life.

Follow-up interviews were sought only with those fathers who had been unemployed at the time of the original interview—a total of 24 men. Personal follow-up interviews were conducted with 22 of these men; telephone contact was made with one father who preferred not to do the personal interview, but agreed to an abbreviated telephone interview; in the last case, the respondent could not be located—the interviewer was informed that he had separated from his former partner and changed his place of residence. Descriptions of the follow-up survey are based on the sample of 22 cases for which the interview was complete.

At the time of the follow-up, nine of the men were still unemployed. Thirteen were employed, eleven of them on a full-time basis.

Comparing the descriptions of the children of the unemployed and the recently employed fathers, there is a tendency—though not statistically significant—for the descriptions of the still-unemployed fathers to be more negative than those of the employed fathers. (T-tests were used to compare the mean descriptions for the two groups of fathers.)

Given the small size of the subsample of still-unemployed fathers from the follow-up survey, it is more appropriate to examine the qualitative survey data than to search for statistical results. The survey asked still-unemployed fathers a number of open-ended questions; the fathers' answers describe some of the costs and the consequences of long-term paternal unemployment. Since it was the second interview, and since they were already familiar with the project, the respondents tended to be relatively frank and open in describing their experiences.

Feelings of frustration, depression, and hopelessness characterize some of these fathers' descriptions of their experiences with prolonged unemployment.

> I don't see chances of being useful. Society has forgotten me. I don't have a sense of control. There's a pointlessness to life. Especially discouraging is the fact that you know it's a country where people have lots of money. It's almost criminal how society looks after some and forgets others.

Others focus on the indignity of unemployment.

> Unemployment is inhumane. My most prevalent feeling is a loss
> of dignity. You start to "look" for work, then "ask" for work, then
> "beg" for work.

The men were asked to think back to the first interview, six
months earlier, and to consider whether things had improved or
gotten worse. None reported that things had gotten better; virtually
all described worsening financial and emotional difficulties. One
father described how child care, which he had found to be rewarding initially, became less rewarding with the passage of time.

> My situation has gotten worse. At first it was good to be at home
> with the kids, but now it's not so good. I should be at work, not
> being a househusband.

At the time of the follow-up interview, one of the still-unemployed fathers was going through a legal separation and child cus-
tody battle. It was his feeling that his unemployment might have
contributed to the break-up of his marriage. This father felt certain
that his situation had deteriorated since the initial interview.

> Things have gotten worse. There's stress due to custody and sep-
> aration procedures. I am unsure of even my immediate future.

Another father described feelings of depression experienced by
himself and his partner.

> We have had to cut down on our recreation. We are both de-
> pressed, and even though I have the time to do things, I am not
> in the mood.

In time, most of these fathers found that the activities of child
care and housekeeping lost their novelty, along with any appeal they
might have had. The following comments are representative:

> It is boring at home doing the housework.

> I don't like keeping house, baby-sitting, changing diapers, and so
> on.

PHASE TWO:
A SURVEY OF UNEMPLOYED FATHERS

This study was designed to explore the pilot study findings using a larger, more systematic, and representative sample. We were particularly interested in whether or not we would replicate the finding that length of unemployment was related to a more negative view of the child. We were also concerned about sampling techniques. In the pilot study we had recruited a sample by asking men with young children whether they would participate. It seemed possible that the reason half of the men had indicated that being with their children was a "positive" aspect of unemployment was due to our recruitment procedure—we were clearly missing those fathers unwilling to be out with their children in the middle of a weekday. Also, the number of fathers interviewed was quite small.

This survey used a sample twice the size of the pilot study sample. Sample design is a key issue in a study of this type. The dilemma is that potential sampling frames such as recipients of unemployment insurance benefits are confidential data and are not accessible by researchers. The 62 respondents in the sample were located through a variety of means, including union and employer referrals from firms experiencing layoffs or plant closures, and direct screening of households via telephone. Because of anticipated difficulties in identifying the target population, the age range of children in the study was expanded. This sample thus consists of unemployed fathers with one or more children 12 years of age or younger. Some of the men have partners who are employed full- or part-time; others have partners who are full-time homemakers. The objective of this study was to examine the impact of paternal unemployment on the lives of these men and their families. Through personal interviews conducted in the men's homes, in the language of their choice, this study used a combination of structured and open-ended questions to elicit the men's feelings about their joblessness.

A subjective definition of unemployment was the basis for inclusion in the sample: Men who considered themselves to be unemployed were eligible. The average duration of unemployment for the sampled men was 8.7 months. Excluded from the sample were men who were voluntarily out of the labor force—so-called "househusbands" who may have opted to stay at home to care for children—as well as those unable to work for health reasons.

The sample represents considerable diversity with regard to education, occupational history, and ethnic background (see Tables 4.1–4.4). The fathers range in age from 20 to 57 years, with an average age of 37.6 years.

The sample includes both men whose spouses are in the labor force (57%) and those whose wives or partners are homemakers (43%). In 30 percent of those families with employed wives, the wives had entered the labor force in response to their husbands' unemployment.

The study examined the impact of paternal unemployment by addressing three sets of research questions: (1) the psychological effects of unemployment on the fathers themselves; (2) the effects on their families; and (3) the effects on their role as fathers.

Effects of Unemployment on the Fathers

> It's goddamn boring staying home. It drives you nuts. The days seem like weeks. In summer you can take the kids outside, or work on the house. Winter is the worst.

> It's frustrating taking care of kids. I'm not used to it. Most of my life I've spent working. It's hard to get used to not working.

> Being unemployed is rotten. I feel useless. I have a family to support and can't find a job. It's heartbreaking. I fill out applications and they don't want to talk to me at all.

Unemployment and the Work Ethic.   The unemployed fathers displayed a strong work ethic. For the most part, when they were working, they liked their work. When asked whether they had found their previous job satisfying, 70 percent replied that they had. Some were very enthusiastic in their descriptions of those jobs, describing skills they had acquired and other benefits they had received. One man who had worked as a meat cutter reports:

> I liked the job. My technique kept improving, and I was given more responsibility. I felt I had to do it right, and I felt worth.

An unemployed truck driver describes that job very positively, but is pessimistic about future job opportunities.

> I really liked that job. The money was good. I don't think I'll ever get a job like that again.

Others reported that their jobs had been "rewarding," "interesting," or "enjoyable."

TABLE 4.1   Respondents' Education Level (n = 60)

	%	n
Elementary School	25.0	15
Some High School	31.7	19
Completed High School	21.7	13
Community College	6.6	4
University	8.4	5
Post Graduate and Professional	6.6	4
Total	100.0%	60

TABLE 4.2   Respondents' Occupation in Most Recent Job (n = 62)

	%	n
Professional/Technical	9.7	6
Managerial/Supervisory	6.5	4
Clerical	4.8	3
Sales	4.8	3
Service	14.5	9
Skilled Trades	22.6	14
Semi-skilled and Unskilled	37.1	23
Total	100.0%	62

In general, the men were extremely anxious to return to the labor force, and many reported that they would consider working at different kinds of jobs in order to do so. A number expressed their personal difficulties accepting unemployment insurance benefits, saying they would much prefer to earn a salary than to receive such benefits. Fully one-half of the men reported that they would accept

TABLE 4.3 Respondents' Ethnic Backgrounds (n = 62)

	%	n
Canadian	29.0	18
Portuguese	11.3	7
East European	9.7	6
Greek	8.2	5
Italian	3.2	2
Chinese	4.8	3
Other Far East Asian	12.9	6
West Asian	9.7	6
West Indian	3.2	2
Spanish	1.6	1
British	1.6	1
Other	4.8	3
Total	100.0%	62

TABLE 4.4 Primary Language Spoken in Respondents' Homes (n = 62)

	%	n
English	61.3	38
Portuguese	11.3	7
Greek	6.5	4
Chinese	4.8	3
Polish	3.2	2
Spanish	1.6	1
Italian	1.6	1
Other	9.7	6
Total	100.0%	62

a job that paid two-thirds of their previous earnings, though many reported that it would be difficult to provide for their families with a pay reduction of such magnitude. Explaining why he would accept a job offer with such a pay cut, one father echoed the sentiments of many.

> Yes, I would accept the job, for I believe it is better to make a little than to make nothing.

For many of the fathers interviewed, receiving unemployment insurance benefits is equivalent to "making nothing."

Most of these men experienced unemployment as a serious blow. Asked to describe the experience of unemployment, 90 percent of the men used predominantly negative terms.

Boredom and Depression. It is boredom that most of the unemployed fathers find to be among the most punishing aspects of their situation. While many claim to take considerable responsibility for household and child care tasks, many describe their unemployment and time spent at home as "doing nothing." One father expresses the sentiments of many.

> I hate it. When you're used to being busy, doing nothing at home is frustrating and depressing.

Boredom, depression, and idleness predominated in the fathers' list of complaints. When asked to describe things they particularly disliked about being unemployed, well over one-third of the fathers (38.7%) began with boredom and related factors.

Asked to indicate their agreement or disagreement with the statement: "Since I have been unemployed, I have often felt bored," the great majority of fathers (88.6%) noted agreement. Some 60 percent of fathers noted that they agreed strongly with the statement, indicating the depth of their feelings.

"Frustrating," "depressing," and "degrading" are other words used by the fathers to describe their unemployment. Over three-quarters of the fathers interviewed (78.8%) reported that they had often felt depressed since becoming unemployed. Still more (88.7%) stated that they had often felt frustrated.

Men's feelings of depression were assessed through a set of statements dealing with feelings about unemployment. For each statement, respondents used a seven-point scale to indicate their agreement or disagreement. Factor analysis of the responses identified a depression factor. Analysis of the fathers' scores on this measure indicates a tendency for the men's feelings of depression to increase with increasing length of unemployment. The depression factor showed a correlation of .26 (p < .05) with the duration of unemployment, measured in months. The words of one unemployed father dramatize this statistical relationship.

> I could tolerate unemployment for two or three months, but it's unbearable now.

For some of the men, the experience of repeated rejection caused the most pain. For others, worry about family finances was the key source of strain. For most, the combination of boredom, financial uncertainty, and lack of future job prospects creates acute emotional strain.

> I've never experienced anything like it in my life. I was brought up to work. It's been very hard on me emotionally, and it's been very hard on the family. You feel useless and bored. After time, you lose confidence. It's frustrating.

Why Me?   The survey also inquired about the men's feelings about the reasons for their unemployed status. The interview asked: "What do you feel is preventing you from finding a job?" The majority of respondents (82.2%) replied in terms of structural conditions, including current economic conditions, rates of unemployment, or problems in particular industries. "It's mainly the economic situation—times are tough." Some cited competition for scarce jobs from immigrants and from women; others blamed automation.

Another group, a minority of the respondents (17.7%), offered more personal explanations for their lack of success in the job market. Some of these men felt they were victims of discrimination because of race or age.

> The problem is my age. Tell them you're 49, they look at you and laugh. If I were 10 years younger, it would make a difference.

> The major reason is that I'm Black. They say the reason is that I don't have relevant Canadian experience.

Others felt their lack of fluency in English hampered their chances of finding work. Still others blamed their situation on their lack of education, training, or experience.

The interview also probed men's feelings of optimism or pessimism about their future employment opportunities. The men were asked: "What do you think about your future job prospects?"

While most indicated that they were generally hopeful about finding work eventually, about one-fifth of the men (21%) sounded hopeless and were pessimistic about the future.

Health Effects.   Some of the fathers felt that the strain of their unemployment had adverse consequences for their physical health—and in some cases for the health of members of their family. Inactivity and lack of schedule or structure in their days made some of the fathers put on weight. Seven reported that they had gained weight since becoming unemployed. Several noted that their consumption of alcohol had increased since they lost their jobs.

Asked: "Have you or any member of your family experienced any change in physical health since you became unemployed?" some 20 percent reported negative health effects that they attributed to unemployment.

Another item in the interview asked for respondents' level of agreement with the statement: "Since I have been unemployed, I have had fewer health problems." Almost one-half (49.2%) disagreed; 30 percent disagreed strongly.

Effects on Families

Interviews with the fathers revealed that most felt that their unemployment had taxed their relationship with their partner—and in some cases with other family members.

> Our marriage is under a lot of strain and we do not get along as well as we would if I had an income. My relationships with my parents and my mother-in-law and father-in-law are under strain as well. They don't complain, but I feel like I'm unproductive and they are not proud of what I do. They see no benefit in my raising the children.

The fathers tend to feel that their job loss has had negative consequences for their partner. This is evident from fathers' responses to the statement: "On balance, my unemployment has been a good experience for my wife/partner." When fathers registered their level of agreement/disagreement with this statement on a seven-point scale, it was apparent that most disagreed. Over half (60.1%) indicated disagreement, with one-half of the fathers choosing the extreme end of the scale to indicate their very strong disagreement with the statement. While some of the fathers agreed with the statement that unemployment had been good for their partner, these were a minority of under 15 percent. About one-quarter (24.2%) chose the neutral point on the scale, indicating neither agreement nor dis-agreement.

While two-thirds of the fathers denied that their unemploy-ment had led to marital problems, almost one-third (31.1%) did feel that their unemployment was responsible for problems in their marriage.

Financial Hardship and Social Deprivation.   Most of the families of the unemployed fathers experienced a drastic reduction in living standard as a result of the father's job loss. The drop in income, in many cases combined with loss of employment-related benefit pro-grams, can mean major changes in quality of life for a family.

The survey compared annual family income levels before and after the father's unemployment. The majority of the families expe-rienced a reduction in total family income as a result of paternal un-employment. About four-fifths of the fathers (81.4%) stated that their unemployment was a financial strain. The drop in the families' an-nual incomes ranged from approximately $5000 to approximately $40,000, with three-quarters of the sample (74.7%) experiencing a drop of between $5000 and $15,000 per year.

Before unemployment, the families in this sample were, for the most part, living at modest levels, with almost 60 percent having to-tal family incomes of less than $30,000, and 85 percent having total incomes below $40,000. The mean, or average, total family income for these families before unemployment was just over $30,000. For many, the loss of income means, first, depleting the family's sav-ings and, subsequently, making changes in housing, recreation, and use of dental and other services.

One of the most sinister consequences of unemployment is the social deprivation that a family may experience. Recreational and leisure activities for children and parents are often reported to be curtailed as a result of paternal job loss. Travel, vacations, family outings to restaurants, and even movies are not affordable for most of the families of the unemployed fathers. The families miss such recreational activities, and a number of the fathers expressed their regret about these missed opportunities.

It doesn't work out well as far as our personal life is concerned.
We haven't had a holiday or even eaten out together in years.

> We've had to draw the line on social things. We're just not as free
> with our money right now—we can only afford the necessities.

One father made it quite clear that the financial difficulty is the
major difficulty they face.

> Being unemployed is not the problem—the monetary thing is the
> problem, the real difference in the monetary situation.

Some of the families are forced to change their housing accom-
modations—or to alter their future plans for housing—as a result of
the father's joblessness. A number expressed concern about being
able to meet their mortgage payments; others had already put their
homes up for sale or had sold their homes.

Their financial hardship associated with unemployment is clearly
a major concern of the fathers. When asked to describe any things
they particularly disliked about being unemployed, financial difficulty was the concern mentioned first by the second greatest number of men, with 30.6 percent citing this as their first response.
Financial hardship ranked second in concern following boredom and
depression. Many other fathers, while first citing some other factor
such as boredom or loss of self-esteem, mentioned the financial strain
as an additional problem.

The fathers' comments revealed the strength of their concerns
about not being able to provide for their families.

> I dislike having to scrounge to make money. It's embarrassing. I
> worry about bills and everything. Where is the grocery money
> coming from?

> I have no money to buy what my family wants. I have no extra
> cash for going out. I can't buy new clothes and other things my
> children want.

The survey identified dental care as an area where a number of
families of the unemployed experienced a reduction in service. Almost half of the fathers (46.4%) reported that their family's dental
care had been negatively affected by their unemployment. For many
of the fathers, dental plans had been included in the fringe benefits
provided by their former employers. Along with job loss, they find
themselves without coverage for dental care.

The fathers' statements indicate that school age children's dental care can be provided through the schools, and it is the parents
who seem to have the greatest need. Preventive dental care is a lux-
ury some of the families can no longer afford.

Not all of the families live in financial hardship as a result of the father's unemployment. A small number (9%) experienced either no change or an increase in income. These were mostly families in which an unemployed father was collecting unemployment insurance benefits and in which the spouse entered the labor force as a result of her husband's unemployment. In one such case, the family's access to dental care actually improved.

Effects on Fathering

The survey asked the fathers to describe any things they particularly *liked* about being unemployed. "Nothing" was the most popular answer, given by almost half (46.8%) of the fathers. ("There is nothing I really like about being unemployed. What could be good about being unemployed?") Lower in frequency than "nothing," but higher in frequency than "a chance to do odd jobs and leisure activities" (17.7%), was the answer "time with kids and family," which was mentioned by almost one-quarter of the fathers (24.2%). While few of the fathers citing this family-related advantage of unemployment were positive in their overall assessment of the experience of unemployment, these men did appreciate that their jobless schedules permitted increased opportunities to spend time with their young children.

In reviewing their answers, it is clear that even those who appreciated the opportunity for increased contact with their children found the experience of unemployment considerably more bitter than sweet. Typical of these answers are the following:

> There are no real good points to being unemployed, but spending more time with my daughter is better than before.
>
> The only good thing is spending more time with my son.
>
> I enjoy only being with my family, enjoying them.
>
> The only thing I like is having more time with my children.

Fathers as Caregivers. To what extent do unemployed fathers assume an active role in child care? Do the fathers take on primary responsibility for child care while they are unemployed? How well-prepared do the fathers feel for the role of child care provider? What about the mothers in these families—how do they influence their partners' involvement in child care and other household responsibilities?

Because it is difficult to gauge the degree to which fathers assume responsibility for child care, the survey used two different indicators. For a quantitative measure of paternal participation, fathers were asked to indicate the total number of hours of care they had provided for the target child in the week immediately preceding the interview. As a more general measure of paternal participation, an open-ended question asked the fathers to describe their predominant child care arrangement for the target child. In addition, fathers' attitudes toward child care activities were also assessed through a number of structured questions.

With regard to the number of hours of child care, there is a considerable range in the amount of child care provided by the fathers (see Table 4.5). A relatively small number (15%) do not provide care for their children. One-third of the fathers provide between 1 and 19 hours of child care weekly. Over one-half of the fathers (51.7%) provide care for at least twenty hours per week. The mean, or average, number of hours of child care provided by the fathers is 23.7 hours per week. The second measure of paternal participation in child care was taken from the fathers' descriptions of the predominant child care arrangement used for the target child. These descriptions were coded according to whether it was the father, the mother, both parents jointly, or some other child care provider(s) who had the primary responsibility. Fathers were designated "highly participant" if they were identified as either the sole or the joint provider of care for the target child.

In 10 of the families (16.4%) it was reported that the father had major responsibility for child care; in an additional 13 families (21.3%) the two parents assumed joint responsibility. The remaining families used a variety of child care arrangements, including the mother providing care (24.6%), relatives, baby-sitters, day care centers, and schools.

Not surprisingly, the two measures of paternal participation in child care are highly related. Fathers who were designated as "highly participant" put in, on average, a significantly higher number of hours of child care per week than did the other fathers. Compared with the mean (average) number of hours of child care of 23.7 per week for the sample as a whole, the highly participant fathers averaged 32.3 hours weekly; the other fathers averaged only 18.3 hours.

The key factor influencing fathers' participation in child care appears to be the employment status of the partner. Men whose wives were employed devoted significantly more hours per week to child care than did men whose wives were full-time homemakers.

TABLE 4.5   Weekly Hours of Child Care Provided by Men with Employed and Homemaking Partners (n = 60)

Hours of Care	%	n
0	15.0	9
1 - 9	11.7	7
10 - 19	21.6	13
20 - 29	15.0	9
30 - 39	13.3	8
40 or more	23.4	14
Total	100.0%	60

As shown in Table 4.6, the mean number of weekly child care hours for men whose wives were employed was 28.3; the men whose wives were homemakers spent an average of 17.2 hours doing child care per week [t(58) = 2.17, p < .05]. In addition, fathers with working wives had fewer difficulties dealing with their children (looked at in terms of one of the factors coming out of the factor analysis—t(59) = 2.06, p < .05). This factor was also directly related to the number of hours fathers spent caring for the child [r(60) = .20, p < .05]. That is, the greater the number of hours of child care, the more likely it was that fathers had fewer difficulties dealing with the children.

It is interesting to note the number of variables that do *not* appear to influence the unemployed fathers' participation in child care. Among those variables are educational levels of father and spouse, family income level before and during unemployment, the father's level of satisfaction with his previous job, the father's age, the sex of the target child, the number of children in the family, and the duration of the father's unemployment.

Supports for Child Care.   Before becoming parents, most North American men have little information, training, or experience in the care of young children. While some such preparation is generally part of the socialization of females, this is often not the case for men. When faced with the responsibility for child care, where do the men turn for information and advice?

TABLE 4.6   Mean Hours of Child Care Provided by Men with Employed and
Homemaking Partners. (n=60)

Partner's Employment Status	Mean Hours of Care Weekly	Standard Deviation	n
Employed	28.3	24.0	35
Homemaker	17.2	15.0	25
		p < .05	

While other researchers have not dealt with this issue as it relates to unemployed fathers in particular, there is one recent American study that looks at fathers in general. Riley and Cochran (1985)
investigated fathers' use of personal social networks as a source of
child rearing advice. They found that 27 percent of the fathers reported no such source of advice on child rearing.

How does this finding compare with the sources of information
and advice used by the Canadian sample of unemployed fathers?
Asked "To whom have you gone for information and advice on child
care?" nearly three-quarters of the sample replied "No one" (see Table 4.7). Another 10 percent reported that they relied on their
spouse/partner for such information; about 5 percent cited professionals such as physicians or teachers; and the remaining 10 percent

TABLE 4.7   Sources of Fathers' Information on Child Care (n = 61)

Source	%	n
No one	73.7	45
Spouse/partner	9.8	6
Professionals	6.6	4
Friends	3.3	2
Others	6.6	4
Total	100.0%	61

relied on friends or other sources. These results indicate that the unemployed fathers have few sources of information and advice on child care.

The survey also assessed fathers' attitudes about the kinds of services that would be of value to them. The men were read a list of ten different services, and asked to rate each on a seven-point scale ranging from "extremely helpful" to "not at all helpful." The collection of services included some directed at the men's needs, for example, employment counselling and support group for unemployed fathers, and some programs for children, for example, short-term day care or after school recreation programs. It is of interest to note that those programs that the fathers rated as most helpful were services that provided child care. The three programs that received the highest proportion of "extremely helpful" ratings included: (1) a preschool children's day care center (44.1% rated it highest); (2) after school programs for school age children (43.1%); and (3) short-term or "drop-in" day care (35.6%).

CONCLUSION

The results of this study demonstrate the close interrelationship between the domains of work and family. Job loss represents an extreme example of major change in the work role. For the 62 fathers interviewed in this study, job loss has had a profound impact on their feelings of self-esteem and their roles and relationships within the family.

Viewed from this perspective, it is clear that there are heavy social costs associated with high rates of unemployment. The research evidence on the interdependence between work and family roles directs our attention to the realm of social and employment policy. Policies known as "job redistribution," with expanded opportunities for part-time work and temporary leaves of absence from the labor force, can increase the options for parents such as those in this study sample. On the one hand, improved opportunities for parental leave can help to remove the stigma of and legitimize staying at home to care for young children. On the other hand, by increasing part-time work opportunities, there may be fewer people who, like the fathers in the sample, end up losers in the job lottery.

Although the fathers in this study were uniformly negative about their unemployment, it did not seem as if this attitude carried over to their children. If there were any positive consequences of unemployment at all, they were increased opportunities to spend time with children and family. Most fathers felt competent about their parenting and positive about their children. The results of the larger survey did not replicate the finding in the pilot study that length of unemployment was related to a more negative perception of the children. In fact, there was a clear indication that those fathers who spent more time caring for their children found their parenting role easier than those spending less time.

These results are consistent with Ray and McLoyd's (1986) review of "fathers in hard times" in which they pointed out that even though many studies have established a link between unemployment and child abuse, unemployment does not necessarily have negative effects on the father–child relationship. They also suggested that the marital relationship and the previous father–child relationship might be important mediators of post-unemployment effects.

The main study reported here involved only one point in time and only fathers were interviewed, limiting the extent to which the results can be generalized. Also, given the large number of dependent and independent variables, and the relatively small sample size, descriptive rather than analytic and multivariate analyses seemed most appropriate.

Future research in this area should overcome some of these limitations. In looking at the relations between the domains of work and family, all family members should ideally be included. Spouses and children should be interviewed, and developmental outcomes should, when possible, be measured. Larger sample sizes would allow for the analysis of interactive effects.

The current study focussed on paternal unemployment. Other researchers have frequently focussed on the effects of maternal employment (Hoffman, 1984). To fully explore the relation between work and family life, the joint employment status of both parents should be a focus of study. This would ideally include an examination of the effects of a large number of possible work arrangements (full-time, part-time, shift work, homemaking, etc.) on the entire family. An important question is the way in which the combinations of different work schedules affect parent–child and marital relationships.

REFERENCES

Barnett, R. C. & Baruch, G. (1986). Determinants of fathers' participation in family work. Working Paper No. 136, Wellesley College Center for Research on Women.

Binns, D. & Mars, G. (1984). Family, community and unemployment: A study in change. *Sociological Review, 32*(4), 662–95.

Brenner, M. H. (1973). *Mental illness and the economy*. Cambridge, MA: Harvard University Press.

Burke, R. J. (1985). Comparison of experiences of men and women following a plant shutdown. *Psychological Reports, 57*, 59–66.

Catalano, R.; Dooley, D.; & Jackson, R. L. (1981). Economic predictors of admissions to mental health facilities in a non-metropolitan community. *Journal of Health and Social Behaviour, 22*, 284–97.

Catalano, R.; Dooley, D.; & Jackson, R. L. (1985). Economic antecedents of help seeking: Reformulation of the time-series tests. *Journal of Health and Social Behaviour, 26*(2), 141–52.

Dooley, D. & Catalano, R. (1980). Economic change as a cause of behavioural disorder. *Psychological Bulletin, 87*, 450–68.

Eisenberg, P. & Lazarsfeld, P. F. (1938). The psychological effects of unemployment. *Psychological Bulletin, 35*, 358–90.

Fagin, L. (1981). Unemployment and health in families. London: Department of Health and Social Security.

Fagin, L. & Little, M. (1984). *The forsaken families*. Middlesex: Penguin Books Ltd.

Hakim, C. (1982). The social consequences of high unemployment. *Journal of Social Policy, 11*(4), 433–67.

Hepworth, S. (1980). Moderating factors of the psychological impact of unemployment. *Journal of Occupational Psychology, 53*, 139–45.

Hoffman, L. W. (1984). Work, family, and the socialization of the child. *Parke Review of Child Development Research, 7*, 223–82.

Jahoda, M. (1979). The impact of unemployment in the thirties and seventies. *British Psychological Society Bulletin, 32*, 309–14.

Johnson, L. C. & Abramovitch, R. (1985). *Unemployed fathers: Parenting in a changing labour market*. Toronto: Social Planning Council.

Levine, J. A.; Pleck, J. H.; & Lamb, M. E. (1983). The fatherhood project. In M. E. Lamb and A. Sagi (Eds.), *Fatherhood and family policy* (pp. 101–02). Hillsdale, NJ: Lawrence Erlbaum Associates.

Liem, R. & Liem, J. (1978). Social class and mental illness reconsidered: The role of economic stress and social support. *Journal of Health and Social Behavior, 19*, 139–56.

Liem, R. & Rayman, P. (1982). Health and social costs of unemployment. *American Psychologist, 37*, 1116–23.

Liem, R. & Rayman, P. (1984). Perspectives on unemployment, mental health, and social policy. *International Journal of Mental Health, 13* (1–2), 3–17.

Margolis, L. & Farran, D. (1984). Unemployment and children. *International Journal of Mental Health, 13*(2), 107–24.

Margolis, L. & Farran, D. (1985). Economic policy and children: The consequences for children of paternal job loss. Paper presented at the meeting of the Society for Research in Child Development, Toronto.

Moen, P. (1983). Unemployment, public policy and families: Forecasts for the 1980s. *Journal of Marriage and the Family, 45,* 751–60.

Parsons, T. (1955). Family structure and the socialization of the child. In T. Parsons and R. F. Bales (Eds.), *Family, socialization, and interaction processes.* Glencoe, IL: Free Press.

Pleck, J. H. (1982). Husbands' and wives' family work, paid work, and adjustment. Working Paper No. 95, Wellesley College Center for Research on Women.

Powell, D. W. & Driscoll, P. F. (1978). Middle class professionals face unemployment. *Society, 10,* 18–26.

Ray, S. A. & McLoyd, V. C. (1986). Fathers in hard times: The impact of unemployment and poverty on paternal and marital relations. In M. E. Lamb (Ed.), *The father's role: Applied perspectives.* New York: John Wiley.

Riley, D. & Cochran, M. (1985). Naturally occurring childrearing advice for fathers: Utilization of the personal social network. *Journal of Marriage and the Family, 47*(2), 275–86.

Schwefel, D.; John, J.; Potthoff, P.; & Hechler, A. (1984). Unemployment and mental health: Perspectives from the Federal Republic of Germany. *International Journal of Mental Health, 13*(1–2), 35–50.

South, S. J. (1985). Economic conditions and the divorce rate: A time-series analysis of the postwar United States. *Journal of Marriage and the Family, 47*(1), 31–41.

Swinburne, P. (1981). The psychological impact of unemployment on managers and professional staff. *Journal of Occupational Psychology, 54,* 47–64.

Thomas, L.; McCabe, E.; & Berry, J. E. (1980). Unemployment and family stress: A reassessment. *Family Relations, 29,* 517–24.

Warr, P. B. (1978). A study of psychological well-being. *British Journal of Psychology, 69,* 111–21.

5

THE POTENTIAL ROLE
OF NATIONAL SURVEYS
AS A TOOL IN UNDERSTANDING
THE ECOLOGY OF CHILD CARE

DONNA S. LERO

Ecological research frequently is associated with intensive and often multiple-instrument means of data collection at microsystem through exosystem levels. The methodological requirements of large-scale national surveys might initially seem to be at odds with an ecological approach, yet Donna Lero convincingly describes a role and approach for national surveys that is supportive of ecological inquiries.

As Eichler (1983) and Kamerman and Hayes (1982) have noted, dramatic increases in married women's labor force participation, especially among mothers of young children; a rapidly rising divorce rate; and a simultaneous sharp decline in the fertility rate in North America are resulting in major changes in the form of the family and in the roles and relationships of its members. "Taken together, these changes in work and family affect virtually all our social, cultural, and economic arrangements. Most importantly, they affect the environments in which children are reared and the experiences they have growing up" (Kamerman & Hayes, 1982, p. 3).

According to studies conducted by Statistics Canada (1982) and the U.S. Bureau of the Census (Lueck, Orr, & O'Connell, 1982), the majority of families with preschool age children now rely on some form of regular supplemental care for their children as a necessary family resource, as do many parents of school age children. Existing formal child care services (licensed group programs and supervised family day care homes) seem to meet the needs of a minority of families in North America. Yet, despite several attempts to do so, such basic questions as where are the children? what factors are the most critical determinants of current use patterns? and what forms of care do parents need and want? remain unanswered. These questions are of most direct importance to governments and policy analysts who, faced with increasing demands for additional services, must develop policies and programs that, ideally, ensure that all families have fair access to affordable, high quality child care arrangements that are compatible with their needs and preferences.

Other questions that are equally pressing have both practical and theoretical importance. They are based on the assumption that each family's work and child care arrangements represent realistic adaptations to a complex set of needs, desires, pressures, and constraints. From an ecological perspective, the use of one or more child care arrangements (including care by one or both parents at home) is construed both as a major dependent variable and as a factor that, in turn, affects children's daily experiences and has both direct and indirect effects on each parent, their relationship to each other and to their children, and their relationship to the world of work.

Since Bronfenbrenner (1979) first suggested that the context of development might best be thought of as concentric ecological systems and their interrelationships, significant efforts have been made to better understand how elements within microsystems and relations between them (mesosystems) affect children's experiences in families and in day care settings. Observational studies of structural and social characteristics and interactions taking place within day care settings have been particularly important for identifying the features of those environments that promote children's cognitive, language, and social development (Fosburg, 1981; Goelman, 1986; McCartney, 1984; McCartney et al., 1982; Rubenstein & Howes, 1979; Ruopp et al., 1979). Additional studies such as those conducted by Powell (1978) and Pence and Goelman (1987) have identified variations in communication patterns and relationships between day care providers and parents. Bronfenbrenner, Alvarez, and Henderson (1984) have provided a fascinating analysis of how work and family variables can interact to result in quite different perceptions of three-year-old sons and daughters by their working parents.

Other studies have attempted to investigate the determinants of parents' preferences among child care arrangements, their search and decision-making processes, and actual use patterns (Johnson, 1977; Lein, 1979; Lero, 1981; Michelson, 1983; Pence & Goelman, 1987; Powell & Eisenstadt, 1983; Rutman & Chommie, 1973). Earlier beliefs that preferences or choices among day care alternatives could be predicted from a limited number of socio-demographic variables, coupled with the child's age, have been refuted as researchers have cast their nets over broader domains. Instead, it seems that child care choices are affected by a wide range of child, family, employment, and community characteristics—and that prediction equations may be different for different subgroups (i.e., the person-process-context model applies). Not surprisingly, inconsistencies have emerged in the findings, as one might expect when one pauses to consider that studies of the determinants of parental preferences and child care use patterns are "ecologically bound." That is, the results are likely to be strongly affected by socio-demographic and ethnic differences among samples, the nature of parents' employment opportunities, and the number and nature of child care options that exist in given locales. Taken together, these observational and smaller-scale survey studies help provide clues to the complexity of variables, processes, and interactions at work in the real world of families' and children's experiences.

NATIONAL CHILD CARE SURVEYS

The purpose of this chapter is to identify the potential role of national child care surveys as a complementary research approach to experiments and smaller-scale studies in adding to our understanding of the ecology of child care. The purposes and characteristics of national household surveys will be discussed as a means of identifying the opportunities and obstacles they provide for studying the interactions among work, family, and child care systems. A particular survey that has been proposed to enable an ecological investigation of child care patterns in Canada (Lero, Pence, Goelman, & Brockman, 1985a) will form the basis of much of the final section of this chapter. That section sets out the approach and parameters that are required to make maximal use of the national survey as a tool for understanding the relationships among work, family, and child care microsystems within even broader social, geographic, and service delivery contexts.

Previous National Surveys

Seven national surveys of families and their child care arrangements have been conducted in the United States and Canada to date. They consist of surveys conducted by the U.S. Bureau of the Census in 1958, 1965, 1977, and 1982, in which data were collected as part of or a brief supplement to the Current Population Survey (CPS), and two surveys conducted by Statistics Canada as supplements to their ongoing Labour Force Studies (LFS) (1973, 1982). An additional survey conducted by Rhodes and Moore in 1975 (National Child Care Consumer Survey) is the only survey among the seven that was conducted by private researchers with funds provided by the U.S. Department of Health, Education and Welfare. Individually and together, these studies have provided both interesting descriptive data about the numbers and percentages of children in various types of care and a few noticeable trends. The studies have not, however, made significant contributions to our understanding of ecological relationships. The reasons for this failure relate in part to limited purposes and resources, but they also reflect the fact that the ecological approach is still fairly new, and the volume and complexity of variables and relationships within it present many challenges.

Purposes for Which National Surveys Are Used

To appreciate both the opportunities and the obstacles attached to national surveys, it is important to understand the purposes for which they are usually employed, the conditions and methodological constraints that may apply, and the factors that can limit theoretical, scholarly data analyses.

National surveys are used primarily to furnish precise estimates of the number and distribution of families or children using various child care arrangements. The population for which estimates are made typically includes all civilian non-institutionalized women of a specified age range with at least one child under a specified age (e.g., five years), or all civilian households (with certain exceptions) containing at least one child under the age specified. In some cases the estimates apply only to women who were employed at the time of the survey or during the 12 months preceding it (U.S. Bureau of the Census, 1977; Statistics Canada, 1973). In every case (except the Rhodes & Moore study), national surveys have been conducted by a government bureau charged with the rigorous collection of national data in the public interest. In large part, these surveys were commissioned by the Canadian and U.S. governments for the purpose of establishing data bases for internal policy-related purposes. In addition to demographic questions, the surveys typically contain questions about child care arrangements used while parents (or mothers) are working, questions about payments made for child care, and possibly one or two questions about the potential effects of limited availability of child care services on mothers' labor force participation.

It is worth noting that all of the government surveys have been conducted in conjunction with the collection of labor force data, in part due to the obvious relationship between parental employment and child care needs. It is also cost-efficient and convenient to use an existing mechanism that provides a nationally representative sample of households for which useful socio-demographic and employment data are available. A by-product of this approach that should be recognized, however, is that data on child care use patterns and government policies continue to be based almost exclusively on parental employment—especially maternal employment characteristics. Supplemental child care has generally not been appreciated by demographers and policy analysts as a family support used for a variety of purposes besides providing care while mothers or both parents are at work. A typology of the broader range of purposes for which supplemental child care is used by parents appears in Table 5.1.

As a means of providing reliable and valid national statistics about the incidence and prevalence of child care arrangements, national surveys cannot be matched. They provide precise estimates based on responses given at a particular point in time by a large, statistically representative sample of families with children. Hence, they are ideal for answering questions such as:

> What proportion of children three–five years of age attend a day care center for thirty or more hours per week?
> How many children age six–twelve are "latch-key children," providing care for themselves on a regular basis while their parents are at work?
> What proportion of dual-earner families use a nonrelative in or outside their home to provide care for children under the age of five?
> What proportion of families utilize a combination of two or more care arrangements for their youngest child in a usual week? What combinations are most common?
> How large is the proportion of households in which extra-parental care is required for two or more children below the age of seven?

Providing reliable answers to such questions serves several purposes. First, it provides an overall sense of how children are cared for. In essence, the results provide researchers, politicians, and citizens with a reading of the national child care pulse. Once obtained, these

TABLE 5.1 Purposes/Functions Served by Supplemental Child Care

1. To provide appropriate care when both parents (or a single parent) are working, or are engaged in an employment-related activity.
   --Full-time employment
   --Part-time employment
   --At a conference or traveling out of town in connection with parental work
   --Engaged in farm labor
   --Looking for work

2. To provide appropriate care when parents are continuing their education, or are enrolled in a training or retraining program.

3. To provide care for children and support for families with special needs. For example:
   --Families in which one or both parents have chronic health problems
   --Families experiencing, or at risk of experiencing, significant distress
   --Families with a background of, or high risk of, child abuse and neglect
   --Families with a handicapped or chronically ill child

4. To provide children with opportunities to participate in experiences designed to stimulate their physical, intellectual, and emotional development; promote personal competence; and enable the development of social skills through interactions with other children and adults.

5. To provide children who have special needs (e.g., retardation, sensory impairments, etc.) with an opportunity to receive specially designed stimulation and remediation in a setting that allows social interaction with other children and adults.

6. To provide appropriate child care as a supportive resource to families at specific times of peak need occasioned by such circumstances as:
   --A family illness or emergency
   --Childbirth
   --Seasonal employment
   --Lack of availability of regular caregivers
   --School or program closings during summer months, professional development days, holidays, etc.

7. To provide appropriate care when parents are engaged in:
   --Volunteer or community activities
   --Religious or ethnic groups/activities
   --Personal or social tasks or activities
   --Family tasks (with, or on behalf of, family members) at which time it is appropriate or desirable to utilize supplemental child care

statistics serve as a reference point in comparison with other countries and with ideal or desired standards. Second, these descriptive data can be used to measure changes in child care patterns over time. A third purpose is policy-related. Specifically, social policy and economic analysts can use such data for the purposes of planning alternative tax treatments and policy directions and of assessing the effects of existing policies. Finally, accurate data on child care use patterns can provide researchers with contextual information and suggestions for future research. As examples, when less than 15 percent of children under the age of five are cared for in licensed settings, researchers should be especially sensitive to selection processes at work that may influence results of program evaluation and comparative studies. The emergence of new or unexpected patterns of care may precipitate research questions and studies that may not have otherwise been considered.

In addition to providing prevalence estimates, national surveys can be used to identify distributional characteristics. In this way, the data reveal how children and families are distributed into child care user groups across various strata. Comparisons of use patterns in different social and economic strata can be used for purposes similar to those already described. They can also suggest some ecological linkages by denoting how variations in configurations of geographic, family, and employment variables relate to child care use. Unfortunately, such contributions have been few in number. In most instances, cross-tabulations of user groups by selected socio-demographic variables (race, income, full-time vs. part-time employment, blue- vs. white-collar occupations) are presented as descriptive data, in character with a structural "social address" model (Bronfenbrenner & Crouter, 1983). These cross-tabulations are interesting, but the findings are limited by the absence of information that might illuminate the processes and relationships behind observed statistical associations. The most recent U.S. survey (June 1982) did include several interesting multiple classification analyses in order to provide "profiles" of particular user groups. All in all, however, national surveys have been conducted for the purpose of providing precise descriptive information, and not for the purpose of building or testing complex theoretical models or exploring the complexity of multivariate and multisystem relationships.

Methodological Features of National Surveys

There are obviously a number of strengths associated with the careful and rigorous way in which national surveys are conducted.

Such surveys, conducted as they are in thousands of households, are expensive undertakings. For this reason, particular attention is given to question wording, the selection of variables, standardizing interview and data coding procedures, and avoiding ambiguities and response burden.

Unfortunately, these efforts to maintain high reliability and minimize the time that will be required to collect the data often can be frustrating to academic researchers, who are likely to find that they must: (1) drastically reduce the number of questions they would like to ask; (2) avoid open-ended questions or procedures that might be too time-consuming or overly expensive to mount; and (3) reduce the number of response alternatives they might prefer. In short, the ecological researcher who is interested in complex patterns of relationships within and between settings may find that in using a national survey she/he must continually struggle to reach the best compromise between depth and breadth, quantity and quality, and efficiency vs. richness of detail.

In addition, the use of a nationally representative sample challenges researchers to make most questions relevant to *all* possible respondents and/or to carefully design questions or sections of interview schedules (and appropriate skip instructions) for selected individuals or groups (e.g., single parents, families with children of specific ages, those using care provided by a relative, etc.). This task is considerably more difficult and complex than developing interview schedules or questionnaires for more select groups, such as parents of preschoolers attending day care centers.

Two additional concerns that must be addressed relate to the use of words and terminology that may not be understood by respondents or may have different meanings in different provinces or regions. The word "kindergarten," for example, refers to publicly funded pre-grade one classes in Ontario; to municipally funded, community-based programs in Alberta; and to privately or cooperatively run preschool programs for five-year-olds in Prince Edward Island and New Brunswick. These programs may be offered on either a half- or full-day basis in different communities. Hence, using the word "kindergarten" without collecting additional information about sponsorship and hours can lead to erroneous conclusions. A related concern that applies is the need to carefully check translations of instruments when several languages are used.

Certainly, one of the hallmarks of national studies is the large sample size that is used. Parenthetically, most government-conducted household surveys generally have high response rates (90% or better), since families, if selected for monthly Labour Force or CPS sampling studies, are required to participate in the collection of labor force data. While supplements are optional, response rates are heightened by their association with the required data collection and by the steps that are taken to minimize the burden on potential respondents by keeping interviews and questionnaires brief, unambiguous, and interesting. Collection of the data by trained interviewers under legislation that ensures anonymity and confidentiality is also helpful. Sample sizes in the American and Canadian studies range from 5000 to 56,000 households. One might naturally assume that such large samples can yield very specific data. Unfortunately, several examples exist in which findings have been severely limited due to small subsamples or to response alternatives that do not discriminate as effectively as possible. As an example, several national surveys present detailed information about child care arrangements for heterogeneous age groups, that is, children ages 0–5 or 6–14 (U.S. Bureau of the Census, 1977, 1982; Statistics Canada, 1982). Clearly, the implications of findings of the prevalence of "latch-key" arrangements are quite different for 6–10-year-olds than for teen-agers. The dynamics of child care use are also obviously quite different for infants than for preschoolers ages 3–5. The ironic fact is that samples of 25,000 families are not large enough to provide fine-grained estimates of the use of many different kinds of arrangements by age of child over a broad age range for different types of families. Reliable estimates of the kind that would be useful for comparative ecological purposes require even larger sample sizes.

Constraints on Data Analysis

Ecological analysis of national survey data is obviously constrained if the purposes of the study are highly specific and do not extend past generating general descriptive data, if the number and range of variables is limited, if the sample size is not large enough to allow effective comparisons among subgroups or sophisticated multivariate analyses, if time and resources for more extensive data analysis are not provided, and if the academic research community cannot obtain access to, or easily manipulate, the data. Unfortunately, most of the constraints itemized above have applied and have served to limit many possibilities for scholarly investigation.

Two other constraints have also been operating. One is the fact that our understanding and appreciation of ecological relationships are relatively new. Second, national studies traditionally have been conducted by demographers and other individuals trained in using quantitative, structural approaches to data analysis for descriptive purposes. Researchers who have been trained and who conduct research in single disciplines are unlikely to adopt the broad, interdisciplinary systems approach that is required for ecological investigations. Indeed, it seems that complex, multisystem studies benefit greatly from the collaborative efforts of a range of individuals who understand the dynamics operating within and across those domains with which they are most familiar, and who are willing to explore new relationships from a systems perspective.

In summary, national household surveys on child care patterns in North America have provided interesting and informative data of a general nature on the incidence of particular types of child care, especially for children younger than five years of age whose mothers are employed in the paid labor force. As yet, however, they have not fulfilled their potential for helping us understand the complex relationships that exist between the worlds of work, family life, and child care. The following portion of this chapter describes a study that has been proposed to the federal and provincial governments in Canada by Lero, Pence, Goelman, and Brockman (1985a, 1985b) in order to begin that process.

CANADIAN FAMILIES
AND THEIR CHILD CARE ARRANGEMENTS:
AN ECOLOGICAL ANALYSIS

Scope, Objectives, and Assumptions

The proposed study has both descriptive and explanatory purposes. On the one hand, it is designed to provide valid, comprehensive information at both the provincial and national levels about families' child care needs, use patterns, preferences, and concerns. On the other, it has been designed to enable ecological analyses of the relationships between family, work, and child care variables in the broader context of social, economic, and geographic factors that impinge on those relationships. A model of the levels and systems that are presumed to be operating is shown in Figure 5.1.

Simply including many variables from different domains, however, is not sufficient to make a survey ecological in its orientation. As Bronfenbrenner (1979) has defined it,

> An ecological experiment is an effort to investigate the progressive accommodation between the growing human organism and its environment through a systematic contrast between two or

FIGURE 5.1   An Ecological Model of Child Care

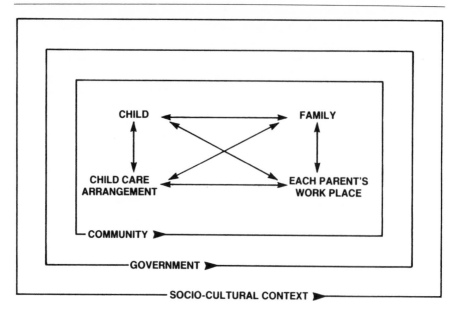

more environmental systems or their structural components, with a careful attempt to control other sources of influence either by random assignment (planned experiment) or by matching (natural experiment). . . . It is from this perspective that the primary purpose of the ecological experiment becomes not hypothesis testing, but discovery—the identification of those systems properties and processes that affect and are affected by the behavior and development of the human being. (pp. 36, 37)

In order to fulfill the ecological mandate that was identified, it was necessary to take the following steps. First, assumptions had to be stated that clearly identified the overall conceptualization behind the study. Second, once specific systems were identified as essential for inclusion in the study, the specific elements within each system (system properties) that would be important to assess had to be identified. Third, the research team had to begin to plan ways that variables within and especially across systems could be conceptualized and measured in order to assess mesosystem relationships and processes operating within the ecological system as a whole. Finally, a temporal dimension was built into the study in order to investigate how characteristics of the family's experiences with child care over a 12-month period for a specified target child in the family affected the child involved, the parents, and their relationships to the world of work. Once these steps were taken, the research team then had to deal with some hard decision making, since it is impractical, if not impossible, for any one research study to encompass all of the variables, levels, and processes that might be conceptually relevant to the broadest ecological perspective. Indeed, it became clear to the research team that additional studies on selected topics could and should meaningfully follow the national survey, using smaller, more select thematic topics and subpopulations. As an example, one topic of interest is child care patterns and concerns in families with a child who has special needs due to health problems, developmental delays, or learning disabilities. A national survey using representative sampling techniques would not yield a large enough number of such families to allow reliable analyses.

The following assumptions underlying the study can now be stated:

1. The choice of a child care arrangement (or a combination of arrangements) is one of several adaptations that families make in response to changing work patterns, perceptions of children's developmental needs, and personal goals within the limits of their own and their community's resources.
2. That choice, at any particular point in time, has other effects on children's development, the family, the community, and society itself—a fact that Kamerman and Hayes (1982) and Belsky, Steinberg, and Walker (1982) have noted has received little attention by social scientists. Furthermore, these effects are likely to be different in single-parent homes, dual career families, and families in which fathers assume a major role in providing child care.
3. Child care use patterns, in addition to reflecting family adaptations to factors internal to the family and the children involved, are affected by the range of alternatives available to families in their communities and by parents' perceptions of those alternatives.

Returning to Figure 5.1, the focus of the study is mesosystem relationships: the linkages among work, family, and child care. The major dependent variables (those most often presumed to reflect adaptations) are parents' (particularly mothers') reports of their family's child care needs, the types of child care currently being used, parental preferences as to employment and child care options, and the perceived effects of child care experiences over the 12 months preceding the interview. The goals of the study then are:

1. To understand the ways in which various parameters of work, family life, characteristics of children, community resources, and social, economic, and geographic factors interact to influence (a) the nature of child care needs that families experience, and (b) parents' preferences for and use of alternative child care arrangements.

2. To examine the impacts of using specific child care arrangements on individuals and families, including (a) problems experienced by families using different child care arrangements; (b) parents' satisfaction with child care arrangements; (c) family tension about meeting child care needs; and (d) effects on parents' workforce participation, absenteeism, and career plans.

Systems and System Properties

The microsystems that are involved in this study are the family, current child care arrangements, and each parent's work involvement. The mesosystem consists of the linkages among these microsystems. The exosystem is represented by characteristics of the community in which each family is located. Macrosystem variables include provincial policies and practices that affect the number and nature of child care services potentially available, and restrictions on the availability of day care subsidies. A partial list of variables is included in Table 5.2. This table describes data collected from families in which the mother is employed in the paid labor force. Alternative questions appropriate to circumstances in which one parent is *not* employed in the paid labor force, focussing on parental support in a community context, are being developed. For the sake of brevity, this section of the chapter includes a limited number of examples of decisions that were made about which variables (system properties) should be included in the analyses and how they will be employed.

Parental Employment. Previous national studies, although firmly tied to employment-related data, have not included a number of variables that are obviously strongly linked to child care needs and may mediate decisions about child care arrangements and the consequences of those decisions. Earlier studies have been confined to assessing employment status (full-time, part-time, unemployed, or not in the paid labor force), number of work hours (often dichotomized into full- and part-time ranges), and occupation (blue-collar vs. white-collar). In addition to these variables, the proposed study includes such variables as the actual hours parents work (i.e., their work schedules), whether their work schedules are consistent from day to day and week to week or are variable, whether variability in work hours is predictable or not, employer flexibility about work scheduling, and an index of employer support of parental functions (including the provision of on-site day care and parental leave policies). These variables will be collected for both parents.

The decision to include questions about work schedules was suggested by research conducted by Pleck, Staines, and Lang (1980); Pleck and Staines (1983); and Nock and Kingston (1984) on the basis of analyses derived from the 1977 Quality of Employment Survey, conducted for the United States Department of Labor by the Survey Research Center at the University of Michigan. Pleck and his colleagues found that work–family conflict is more prevalent among parents than among childless couples and is most often related to excessive work hours, irregular or undesirable work schedules, and physically or psychologically demanding work. Nock and Kingston (1984) went one step further ecologically to begin to juxtapose how husbands' and wives' work hours and work schedules relate to each other. Their concept of the "family work day" includes three dimensions: the total family work time (the combined number of hours a couple works), the length of the work day (the amount of time one or both spouses work), and off-scheduling (the amount of time only one spouse is at work). More recently, Kingston and Nock (1985) found that off-scheduling varies by composition of the nuclear family, with its incidence being considerably higher in families with young children. Another dimension of the family work day (total hours) was noted to correlate significantly with time spent individually by each parent with children and time spent together as a family unit. Data on each parent's actual work schedule in the proposed study will be used in similar ways. In addition, the nature of work hours is important to appreciate since group care arrangements are generally not available in the evenings and on weekends, necessitating the employment of sitters if a spouse or relative is not available.

The employment variables itemized above are likely to affect parents' use of particular child care arrangements. Other employment-related variables were selected as indicators of possible consequences of unstable or poor quality care experienced over the last year. These include questions about whether concerns over child care arrangements caused either parent to leave or refuse a job, reduce work hours, reduce his/her work commitment, worry about

TABLE 5.2   Selected Variables Included in an Ecological Study of Canadian Families and Their Child Care Arrangements

DATA FROM RESPONDENTS[a]

I. Socio-demographic and Contextual Variables
   Age; marital status; education; family income; number and ages of children; presence of health or developmental impairments in children; ethnicity; province; population of city/town; availability of relatives in household, and within 30 miles; perceptions of neighborhood resources and supports to children and families.

II. Employment and Activity Variables for the Mother (and Spouse/Partner)
   A. In the reference week[b]
      Specific occupation; usual work hours; predictability of work hours; overtime worked last week; current participation in educational or training programs; number and distribution of hours spen in educational activities; child care benefits currently available through employment and place of study; hours spent in unpaid volunteer activities or community meetings.
   B. In the last year
      Work and study pattern over the last 12 months.

III. Child Care Arrangements Used During the Reference Week for All Children in the Household < 13 Years Old
   Detailed information regarding school attendance and care received in 11 different types of care, defined in terms of location, type, and relationship of caregiver to child. Includes information on costs and subsidies, availability of receipts for tax purposes.

IV. Additional Information on Main Method of Child Care Utilized for a Target Child[c] During the Reference Week
   Time since starting to use this method; reasons for choice; difficulties experienced in locating desired care; degree of previous acquaintance with caregiver; degree of satisfaction with specific aspects of care; perceived positive and negative effects of this care on child; desired improvements.

---

[a]Respondents are the persons in a household most responsible for making child care arrangements in the family. In most cases, this is the mother. Socio-demographic information is collected about each parent, whenever applicable.

[b]The reference week is the week prior to data collection.

[c]The target child is randomly chosen within age strata from among the children in the household < 13 years of age in order to ensure sufficient detailed information about the child care utilized for infants < 18 months of age, children age 18-36 months, 3-5 years, and 6-12 years.

TABLE 5.2 *(continued)*

V. Child Care History of Target Child Over the Last 12 Months

  A. Actual pattern of arrangements used during the last 12 months, including reasons for changes.

  B. Perceived impact of changes for child.

  C. Difficulties experienced by parents over the last year; reported effects on each parent's work status, hours of work, commitment to job or career, work performance, degree of worry while at work, participation in educational/training programs, school performance.

VI. Current Perceptions of Alternative Child Care Arrangements
Perceived availability affordability, and quality of specific types of child care; most desired type of care and reasons for non-use, when applicable.

VII. Current Level of Work-Family-Child Care Tension
Factors contributing to work-family-child care tension; factors operating to reduce tension.

DATA FROM PROVINCIAL SOURCES
Types of licensed/approved child care available in the province; policies affecting access to formal child care (age limits, service priorities, etc.); number of spaces per capita in licensed centers, and approved family day care homes in major cities for infants, preschoolers, school age children; number of spaces per capita in representative rural areas; availability of kindergarten, junior kindergarten, and after-school programs; provincial policies related to day care subsidies; average costs of selected types of care; other relevant data.

NOTE: This table describes data collected from families in which the mother is employed in the paid labor force. Alternative questions appropriate to circumstances in which one parent is not employed in the paid labor force, focusing on parental support in the community context, are being developed.

SOURCE: D. S. Lero, A. Pence, H. Goelman, and L. Brockman, Canadian families and their child care arrangements: A proposal for a national child care survey (proposal submitted to the Canadian Department of Supplies and Services, 1987).

child care while at work, or miss work because of arrangements breaking down or other problems.

Current Child Care Arrangements. Current child care arrangements are defined as those used at any time in the reference week (the week preceding the study). Characteristics of care that will be collected include the types of care used, defined according to who is the main caregiver and where care is provided, and the sponsorship of care, as well as the mother's main activity (work, study, or personal or family activities) at that time. Additional child care data that will be included can be found in Table 5.2. Data will enable the identification of combinations of care used to cover parental work hours. Information on child care arrangements used in the reference week will be obtained for each child in the family younger than 13 years of age. This is unusual, in that most studies have targeted only one child in each household (the youngest child, as in the CPS study conducted in 1982). The use of a family perspective, however, requires that we investigate and identify whether and how arrangements for one child are related to arrangements for other children in the family. In the 1977 CPS survey, results indicated that 95 percent of mothers who had two children younger than five years of age used the same principal arrangement for both children—but what that arrangement was, was not clear.

In addition to the "snapshot" of care arrangements used during the last week, a number of variables concerning the *main* care arrangement used for a specified target child in each family will be assessed in greater depth. They include the duration of care, parents' reasons for choosing this main method, parents' perceptions of the positive and negative impacts of the current main method of care on the target child, desired changes in the care arrangement that would improve the quality of care, other desired changes, and parents' satisfaction with specific aspects of the care arrangement.

Investigating Relationships
Within and Between Systems

Perhaps one of the most challenging tasks facing the research team has been planning how to study the complex processes and patterns of relationships within the ecological model. In regard to parental employment, data collected about each parent's actual work schedule will be used to derive variables, including the three components of the family work day defined by Nock and Kingston (1984). In addition, we hope to explore the feasibility of assessing the "family child care day"—a new conceptualization that would juxtapose both parents' work hours and the hours when children are in supplemental care arrangements.

Thought has been given to how to assess both direct and indirect effects of systemic influences. As an example, objective data collected from each province will be used to identify the complex differences that exist between provinces in the number and nature of services available, licensing standards, and requirements for day care subsidies. In effect, we have in Canada a naturally occurring experiment but no information on how policy differences affect parents' preferences and actual child care patterns. As examples, Quebec and Alberta have invested significant amounts of funds in after school programs; Saskatchewan will not license group care for children younger than 18 months of age; Nova Scotia will utilize purchase of service agreements (i.e., provide subsidized care) only in nonprofit centers that satisfy a variety of special requirements; and Manitoba is the only province thus far that requires specialized training for licensed family home day care providers and has set a maximum fee that can be charged by day care centers. The typical national survey, as outlined in the earlier section of this chapter, would simply merge all of the findings together. Specific breakdowns by the most populous provinces would be generated; however, since the only data that would normally be collected would be household data (i.e., information from the respondent), there would be no opportunity to assess macrosystem effects.

In our study, we are using *two data sources* (the respondent and each provincial government) in order to make possible the kinds of comparative ecological analyses that are proposed. The impact of provincial policies will be measured directly by comparing actual child care use patterns, for example, the income distribution of families using licensed care arrangements, within each province. In addition, several questions are included in the interview that assess parents' perceptions of the availability, affordability, and quality of different types of care. Hence, we will be able to determine if provincial policies' effects on use patterns are mediated by parental perceptions of care alternatives.

Another example of steps taken to study inter-system relationships is the construction of new variables. In particular, we have developed a scale to assess work–family–child care tension. This concept is an extension of work–family conflict, as defined by Pleck and Staines (1983). In this case, parents are asked how difficult it is for them, on a day-to-day basis, to juggle work, family and child care responsibilities; what factors contribute to the degree of tension they are experiencing; and what factors help reduce tension. Examples of potential contributors to work–family–child care tension include: problems getting to work on time when dropping off the child at his/

her day care setting or picking him/her up on time; feeling that one's employer/work situation is inflexible and/or uncaring about one's role as a parent; and difficulty in scheduling child care with a spouse or partner so that one parent is available to be with the children. Potential tension reducers include having a care provider who is flexible and can accommodate one's needs, having older children help with family and child care responsibilities, and having a spouse who supports one's work involvement. Plans for analysis of mothers' work–family–child care tension include assessing how tension is affected by current child care and employment variables. It is anticipated that work–family–child care tension may be mediated in complex ways by the particular sources of tension that operate in different families and the availability of family and community resources to act as buffers.

Building in a Temporal Dimension

One of the factors that is recognized in Bronfenbrenner's expositions on the ecology of human development is the dynamic nature of relationships. Intra- and inter-system relationships may change over time, both affecting and being affected by the developing human. As yet, the day care literature has provided little information about families' or children's child care histories. Presumably, the nature and sequence of care experiences is becoming an increasingly important factor in children's early experiences.

In this study, the nature and sequence of care experiences over the 12 months preceding the interview will be collected for the target child in the family (a child younger than 13 who is chosen at random). In particular, we are interested in determining the precursors and consequences (for children and parents) of care histories that differ on dimensions of stability, perceived quality, and the incidence of various problems. Are certain types of families more vulnerable either to selecting less satisfactory, less stable care arrangements or to experiencing negative consequences in those circumstances? Are certain kinds of care arrangements more likely to have shorter lives? What are the perceived consequences of multiple care arrangements in a year for young children? How have child care experiences over the last year affected parents' anxiety at work, work commitment, absenteeism, educational plans? What contribution, if any, does the last year's experience history make to mothers' current degree of work–family–child care tension and to current preferences among child care alternatives?

CONCLUSION

This chapter has presented an overview of how national household surveys relevant to child care have been conducted in the past. It has been suggested that, while the surveys have provided interesting descriptive data at the national level, this research strategy has not been used to full effect for the purpose of investigating those ecological relationships that are becoming more common and more important in the lives of children and families. Examples of the decisions, plans, and problems involved in developing a national survey that does focus on ecological processes have been presented. In particular, the unique features of the study outlined in this chapter include obtaining fairly rich data about each of the microsystems of work, family, and child care arrangements; obtaining precise data with respect to two time periods—the reference week and the last 12 months; creating or deriving new variables that assess characteristics of mesosystems (e.g., the family child care day and work–family–child care tension); and obtaining additional data relevant to the study from other sources in order to enable specific comparisons to be made across provinces with different policies and programs in the child care area. Clearly, such a study poses many challenges, but is also likely to yield many rewards for social scientists who have adopted an ecological approach to studying families and their child care arrangements.

It is fortuitous that both the Canadian and U.S. governments are planning their next national child care surveys for the 1980s. In the most ideal circumstances, researchers will be able to conduct ecological surveys of national interest in both countries and gain the opportunity to compare results. Consultations have been occurring between demographic bodies (Statistics Canada and the U.S. Bureau of the Census), researchers, and policy analysts. Major task forces have been conducted in both countries that have raised the issue of day care to a highly visible level. One may hope that political, economic, social, and intellectual influences can provide the kind of ecological environment that will result in the successful conduct of both studies.

REFERENCES

Belsky, J.; Steinberg, L. D.; & Walker, A. (1982). The ecology of day care. In M. Lamb (Ed.), *Non-Traditional families: Parenting and child development* (pp. 71–116). Hillsdale, NJ: Lawrence Erlbaum Associates.

Bronfenbrenner, U. (1979). *The ecology of human development: Experiments by nature and design*. Cambridge, MA: Harvard University Press.

Bronfenbrenner, U.; Alvarez, W. F.; & Henderson, C. R. (1984). Working and watching: Maternal employment status and parents' perceptions of their three-year-old children. *Child Development, 55*, 1362–78.

Bronfenbrenner, U. & Crouter, A. C. (1983). The evolution of environmental models in developmental research. In W. Kessen (Ed.), *History, theories and methods*. Vol. 1 of P. H. Musson (Ed.), *Handbook of child psychology* (4th ed.). New York: John Wiley.

Eichler, M. (1983). *Families in Canada today: Recent changes and their policy consequences*. Toronto: Gage Publishing Limited.

Fosburg, S. (1981). *Family day care in the United States: Summary of the national day care home study*. Washington, DC: U.S. Department of Health and Human Services.

Goelman, H. (1986). *A study of the relationships between structure and process variables in home and day care settings on children's language development*. Paper presented at the University of Victoria Symposium on Ecological Approaches to the Study of Children and Families, Victoria, BC

Johnson, L. C. (1977). *Who cares? Part 1: A report of the Project Child Care survey of parents and their child care arrangements*. Toronto: Community Day Care Coalition and the Social Planning Council of Metropolitan Toronto.

Kamerman, S. B. & Hayes, C. D. (Eds.). (1982). *Families that work: Children in a changing world*. Washington, DC: National Academy Press.

Kingston, P. W. & Nock, S. L. (1985). Consequences of the family work day. *Journal of Marriage and the Family, 47*, 619–29.

Lein, L. (1979). Parental evaluation of child care alternatives. *Urban and Social Review, 12*, 11–16.

Lero, D. S. (1981). *Factors influencing parents' preferences for, and use of alternative child care arrangements for preschool-age children*. Ottawa: Health and Welfare Canada.

Lero, D. S.; Brockman, L.; Pence, A.; & Charlesworth, M. (1985). *Parents' needs, preferences and concerns about child care: Case studies of 336 families*. Ottawa: Secretary of State.

Lero, D. S.; Pence, A.; Goelman, H.; & Brockman, L. (1985a). *Where are the children?—An ecological survey of families and their child care arrangements*. Proposal submitted to the National Department of Health and Welfare, Ottawa.

Lero, D. S.; Pence, A.; Goelman, H.; & Brockman, L. (1985b). *Canadian families and their child care arrangements: An ecological analysis*. Proposal submitted to the Social Sciences and Humanities Research Council of Canada, Ottawa.

Lueck, M.; Orr, A. C.; & O'Connell, M. (1982). *Trends in child care arrangements of working mothers*. (Current Population Reports. Special Studies, P–23, No. 117). Washington, DC: U.S. Department of Commerce.

McCartney, K. (1984). Effects of quality of day care environment on children's language development. *Developmental Psychology*, 2(2), 224–60.

McCartney, K.; Scarr, S.; Phillips, D.; Grajek, S.; & Schwarz, J. C. (1982). Environmental differences among day care centers and their effects on children's development. In E. Zigler & E. Gordon (Eds.), *Day care: Scientific and social policy issues* (pp. 126–151). Boston: Auburn House.

Michelson, W. (1983). *The logistics of maternal employment: Implications for women and their families*. (Child in the City Report No. 18). Toronto: Centre for Urban and Community Studies, University of Toronto.

Nock, S. L. & Kingston, P. W. (1984). The family work day. *Journal of Marriage and the Family*, 46(2), pp. 333–43.

O'Connell, M. & Rogers, C. C. (1983). *Child care arrangements of working mothers: June, 1982*. (Current Population Reports. Special Studies, P–23, No. 129). Washington, DC: U.S. Department of Commerce.

Pence, A. R. & Goelman, H. (1987). Silent partners: The parents of children in three types of day care. *Early Childhood Research Quarterly*, 2(2), 103–18.

Pleck, J. H. & Staines, G. L. (1983). Work schedules and work–family conflict in two-earner couples. In J. Aldous (Ed.), *Two paychecks: Life in dual earner families* (pp. 63–87). Beverly Hills, CA: Sage Publications.

Pleck, J. H.; Staines, G. L.; & Lang, L. (1980). Conflicts between work and family life. *Monthly Labor Review*, March, 29–32.

Powell, D. R. (1978). The interpersonal relationship between parents and caregivers in day care settings. *American Journal of Orthopsychiatry*, 48, 680–89.

Powell, D. R. & Eisenstadt, J. W. (1983). Predictors of help-seeking in an urban setting: The search for child care. *American Journal of Community Psychology*, 11(4), 401–21.

Rhodes, T. W. & Moore, J. C. (1975). *National child care consumer study*. Unco, Inc. Washington, DC: Office of Child Development (DHHS).

Rubenstein, J. L. & Howes, C. (1979). Caregiving and infant behavior in day care and homes. *Developmental Psychology*, 15, 1–24.

Ruopp, R. R.; Travers, J.; Glantz, F.; Coelen, C. (1979). *Children at the center*. Final report of the (U.S.) national day care study, Vol. 1. Cambridge, MA: Abt Books.

Rutman, L. & Chommie, P. (1973). A comparison of families using commercial and subsidized day care spaces. *Child Welfare*, 52(5), 287–97.

Statistics Canada. (1973). *Working mothers and their child care arrangements in Canada, 1973*. Manpower Research and Development Section, Labour Division. Ottawa: Statistics Canada.

Statistics Canada. (1982). *Initial results from the 1981 survey of child care ar-rangements from the labour force*. (Catalog No. 71–001, Vol. 38, No. 8). Ottawa: Statistics Canada.

U.S. Bureau of the Census. (1977). *Trends in child care arrangements of work-ing mothers, 1977*. Current Population Reports, Special Studies Series P-23, No. 117. Washington, DC: U.S. Bureau of the Census.

U.S. Bureau of the Census (1982). *Child care arrangements of working moth-ers: June 1982*. Current Population Reports, Special Studies Series P-23, No. 129. Washington, DC: U.S. Bureau of the Census.

6

THE ECOLOGY
OF ADOLESCENT PREGNANCY
AND PARENTHOOD

MICHAEL E. LAMB

Michael Lamb utilizes the levels of the nested systems model in a creative juxtaposition of national (macro-level) data with personal (micro-level) characteristics of adolescent mothers and fathers from a Salt Lake City study. Utilization of the ecological model not only serves to clarify the critical importance of socio-familial factors in addressing concerns resulting from adolescent pregnancy, but also suggests entry points for ecological interventions.

Adolescent pregnancy may sound like an unlikely topic for inclusion in a volume concerned with the ecological perspective on child development. The topic sounds inappropriate because until remarkably recently, adolescent pregnancy was treated primarily as a biomedical problem. As a focus of study, adolescent pregnancy was "discovered" in the 1950s and 1960s by medical practitioners, primarily obstetricians, who were concerned about the reproductive risks faced by adolescents bearing children of their own (Battaglia, Frazier, & Hellegers, 1963; Israel & Woutersz, 1963). Twenty to thirty years of research since then have made clear that the biomedical risks of adolescent pregnancy were vastly overestimated, even if such august magazines as *Time* persist in overstating them. Certainly, the biomedical risks are higher when girls of 13 and 14 deliver children of their own, but girls in their early teens constitute a very small proportion of the total number of adolescent childbearers. In 1978, for example, 1.2 million adolescents in the U.S. became pregnant; 30,000 (2.5%) were 15 years of age or younger (National Center for Health Statistics, 1981). The majority of pregnant adolescents are aged 16 or older, and they are in many respects optimally suited for childbearing. The major risks that Americans have to be concerned about, at least as far as the biology of reproduction is concerned, are associated with inadequate nutrition, inadequate prenatal care, and multiparity rather than maternal age *per se*. Happily, even the correlation between maternal age and both inadequate prenatal care and nutrition has been reduced dramatically in the last decade or so, largely through the establishment of a number of federally funded adolescent pregnancy assistance programs established and funded by the U.S. government.

This is not to say that adolescent pregnancy and parenthood are without problems. However, the risks associated with adolescent *parenthood* appear more pervasive than the risks of adolescent *pregnancy*; the former risks are largely social in nature and they can be substantial. The goal of this chapter is to show how the risks of teenage child rearing are related to, if not even inherent in, the social circumstances in which these women and their partners find themselves.

Only recently have social scientists shifted their focus from pregnant adolescents to adolescent parents, thereby acknowledging the potential functions of, and the problems faced by, the fathers of children born to adolescent mothers (Elster & Lamb, 1986). Formerly, to the extent that any attention was paid to these young men, they were viewed as irresponsible, predatory males who had surely taken advantage of their young partners and then disavowed responsibility, leaving the young mothers to cope alone with both societal disapproval and the economic costs associated with bearing and rearing their children. As suggested below, this stereotyped picture of the adolescent father is quite as misleading as was the earlier emphasis on the biomedical dangers of adolescent pregnancy. Although fathers certainly can escape parental responsibilities more easily than mothers can, the majority of adolescent fathers apparently do not attempt to do so; many remain committed—idealistically, realistically, or resignedly—to the assumption and fulfillment of their paternal responsibilities.

In this chapter, I describe the personal and social circumstances of young parents in the United States today, drawing on nationally representative data and the results of research conducted in the Salt Lake City by Arthur Elster and myself. To set the stage, I begin by discussing the incidence of and societal perceptions of adolescent pregnancy; these issues provide an important means of understanding the macro-social context of adolescent pregnancy and parenthood. Then I describe the distinctive personal, or micro-level, characteristics of adolescent mothers and their partners, particularly those in our Salt Lake City population. In the third section, the focus shifts from individuals to the social relationships between pregnant adolescents and their male partners. In the fourth section, I present evidence concerning the longevity of and prognosis for these relationships, and then in the fifth section I turn to a discussion of the educational and occupational status of adolescent parents and of adults who began parenting in adolescence. The implications for parenting and child development are then discussed. The chapter ends with a summary of the major insights gained by adopting an ecological view of adolescent parenting.

Because this is an emergent area of research, many of the paths of influence remain to be explored empirically; for the most part, therefore, we can only speculate about the likely effects of socioeconomic circumstances on parental behavior. We believe it is important to first describe the ecology of adolescent parenthood before attempting to evaluate the formative importance of the various factors identified.

DEMOGRAPHY AND CHANGING SOCIAL ATTITUDES

Much has been written about the "epidemic" of teen-age pregnancy in the United States over the last decade, and it is indeed true that adolescent women today are far more likely to become pregnant than their predecessors were. Since 1978, roughly 1.2 million adolescents (women under 20 years of age) have become pregnant each year, and although some 40 percent elect to undergo abortions, roughly 60 percent of the total choose to bring their pregnancies to term (National Center for Health Statistics, 1981, 1984). A large proportion of these choose to keep their children, rather than place them up for adoption, and this represents a dramatic change from the situation that existed a generation ago. Further, of those adolescents who keep their children, only about a third "legitimize" the births, 10 to 20 percent are married at the time of conception, and about half choose to raise their children out of wedlock. The willingness of young women to publicly acknowledge their pregnancies and attempt to raise their children—often alone—reflects the fact that our society has become much more tolerant of adolescent pregnancy, presumably because it now occurs so much more commonly than in the past.

Acceptance of teen-age pregnancy in the United States, however, is neither universal nor unambiguous. Although the United States has a rate of teen pregnancy many times higher than that of any other country in the industrialized world, this has not brought either understanding or acceptance. Most experts attribute the increased incidence of adolescent pregnancy in the United States to the fact that sex education and contraceptive counseling are seldom adequate, apparently because parents and other moral leaders are concerned that the discussion of sex will foster teen-age sexual activity and thus increase the rate of teen pregnancy. The Alan Guttmacher Institute reports in a recent survey that, ironically, precisely the opposite is true: Those countries in Western Europe that provide universal sex education and make contraceptive services available to sexually active individuals of all ages have rates of teen-age parenthood that are substantially lower than our own (Jones, Forrest, Goldman, Henshaw, Lincoln, Rosoff, Westoff, & Wulf, 1985).

These facts alone suggest that it would be unwise to generalize from findings obtained in the United States to findings obtained in other countries in which the macro-social context is substantially different from ours. In addition, we must be cautious about generalizing findings from other eras in American social history, because social attitudes and prejudices have changed so much (Vinovskis, 1986). Generalization is further limited by substantial ethnic and racial differences in the incidence of and individual response to adolescent pregnancy (National Center for Health Statistics, 1981, 1984). Contrary to popular belief, the vast majority of adolescent parents in the United States are Caucasian, although the rates of adolescent pregnancy are higher among blacks. Blacks are less likely to abort electively than whites are, they are much less likely to marry in order to legitimize their offspring, and those who deliver are less likely to place their children up for adoption, choosing instead to raise their children with the help of their families. Caucasian teen-agers are increasingly behaving like their black peers in these respects, but the ethnic differences still exist. To the extent that they reflect subcultural differences in the perception and acceptance of adolescent pregnancy and parenthood, they too limit the extent to which we can generalize from existing data, and describe an important subculturally variable aspect of the ecology of adolescent parenthood. This concern is especially relevant to many of the findings obtained in our Salt Lake City research, as our subjects are Caucasian males and females from urban middle-class backgrounds. It cannot be assumed that similar findings would be obtained in studies of economically disadvantaged blacks or rural youths.

Overall, approximately 4 percent of all births in the United States are known to be fathered by men under 20 years of age (National Center for Health Statistics, 1981). About 30 percent of adolescent mothers have partners who are adolescents themselves, while an additional 18 percent of adolescent fathers (22,000) are the partners of mothers over 20 years of age (National Center for Health Statistics, 1981). Thus the total number of adolescent fathers is substantially smaller than the total number of adolescent mothers, even if we assume that the majority of babies whose fathers' ages are not listed on birth certificates (39% of the births to adolescent mothers) had adolescent fathers (National Center for Health Statistics, 1981). Most of the analyses reported in this chapter concern adolescent mothers and their partners, regardless of the partners' ages. In the research conducted by my colleagues and me (see below), for example, females averaged 17 years of age at delivery (range, 12.5 to 19.0 years), whereas the males averaged 21 years of age, with a range of 14 to 30 years.

PSYCHOLOGICAL PROFILES

In an earlier study involving a sample of middle-class Caucasian adolescents from Colorado, Jessor and Jessor (1975) had shown that teen-age sexual activity was correlated with a number of antisocial behaviors, so that boys and girls who initiated sexual activity earlier than their peers were also more likely to be involved in the use of illicit drugs, to smoke, to use or abuse alcohol, and to manifest a diversity of behavioral problems. Because contraceptive precautions are so seldom taken by sexually active American teen-agers, one might predict that pregnant adolescents are almost randomly drawn from among the sexually active youths, and thus that the Jessors' findings should presage comparable differences between adolescent parents and nonparents. However, attempts to distinguish prospective adolescent parents from their peers, rather than sexually active teens from celibate teens as in the Jessors' case, have been less successful. Pauker (1971), for example, compared teen fathers with their peers on a number of measures of psychological functioning that were included in a test battery administered before any pregnancies occurred; few differences were found. Nakashima and Camp (1984) reported that the attitudes and characteristics of adolescent mothers did not differ depending on whether their partners were younger or older than 20 years of age and that the fathers themselves were more similar than different with respect to IQ and attitudes toward parenting. They differed only in that the older fathers reported less marital conflict and, as would be expected, more mature ego development than the younger fathers did. Rivara, Sweeney, and Henderson (1985) compared teen-age fathers with age-matched peers from comparable black lower-class backgrounds; nonfathers were more likely to see pregnancy as a disruption of future plans for school, employment, and marriage, and were less likely to have mothers who had themselves been teen parents. With respect to age at first intercourse, frequency of intercourse, knowledge of reproduction, and use of contraception, however, the groups did not differ.

In an attempt to obtain a clear description of middle-class Caucasian adolescent mothers and their partners, Arthur Elster and I have been studying the clients served by a comprehensive health care program—The Teen Mother and Child Program—which Elster directs. Over the first four-and-a-half years that the program was functioning, 305 of the pregnant adolescent patients were able to identify the fathers of their offspring; 275 of them were primiparous, and these young women and their partners have been the focus of our research.

The bulk of our findings has been consistent with the tenor of the Jessors' conclusions. The proportion of the young women who had experimented with drugs (30%) was modestly higher than the local (though not national) norms, as were the rates of such usage by their partners (43%). However, the most striking evidence that these were indeed "troubled youths" came from data we obtained from the young women and their partners concerning their contacts with the police and legal authorities. We found that 58 percent of the males and 42 percent of the females had had some formal contact with the police in connection with offenses other than status offenses—that is, offenses such as truancy which would not be offenses if the individuals were older (Elster, Lamb, Peters, Kahn, & Tavare, 1987). Elster and colleagues' data, presented in Table 6.1, show that the majority of the offenses in which these individuals were involved were at least moderately serious. Unfortunately, national cumulative incidence rates are not available, but the incidence rates for youngsters between 17 and 21 years of age should be roughly half of those reported by our informants, and we would expect them to understate their legal histories, if anything.

We also found a significant correlation between the offense histories of the pregnant teen-agers and their partners: Not only were

TABLE 6.1 Type of Offenses (116 Incidents) Committed by 81 Prospective
Fathers in Utah

FBI Crime Category	No.	(%)
1. Major Crimes Against Persons and Property		
Larceny-Theft	23	(20.0)
Burglary	9	(8.0)
Motor vehicle theft	5	(4.0)
Aggravated assault	4	(3.5)
Robbery	4	(3.5)
Homicide	1	(.8)
Forcible rape	1	(.8)
Kidnapping	1	(.8)
2. Less Severe Crimes		
Driving under the influence	18	(15.5)
Drug abuse violations	12	(10.0)
Liquor law violations	6	(5.0)
Disorderly conduct	5	(4.0)
Simple assault	5	(4.0)
Vandalism	4	(3.5)
Drunkenness	4	(3.5)
Possession of weapons	2	(2.0)
Sex offenses	2	(2.0)
Possession of stolen property	1	(.8)
Runaways	1	(.8)
Offenses against family & children	1	(.8)
Other offenses	8	(7.0)

both partners likely to have been offenders or nonoffenders, but
there was also a significant relation between the severity of their
most serious offenses. In other words, offenders tended to mate with
offenders, nonoffenders with nonoffenders. The male offenders came
from lower socioeconomic status (SES) backgrounds and were more
likely than nonoffenders to come from single-parent families, to be
unemployed, and to be school dropouts (see Table 6.2). They were
also more likely than nonoffenders to have been involved in previous pregnancies, to have a history of alcohol abuse, and to have
manifest behavior problems at school. In all, there was a clustering
of problem behaviors among offenders, suggesting psychosocial
maladjustment prior to pregnancy.

TABLE 6.2   Psychosocial Characteristics of Legal Offenders and Non-Offenders

	Offenders (N=98)		Signifi- cance of differ- ence	Non- Offenders (N=93)	
	No.	(%)		No.	(%)
Relationship at Delivery					
Married	43	(44)		53	(57)
Living together/engaged	11	(11)		14	(15)
Dating regularly	20	(21)		6	(6)
Little/no involvement	23	(24)		20	(22)
Employed	51	(54)	*	65	(72)
School Dropout	59	(63)	*	42	(45)
History of Previous Pregnancy	17	(24)	*	11	(13)
Smoking[1]	30	(48)		23	(38)
Drinking[2]	29	45	*	17	(27)
Substance Use[3]	14	(22)		9	(14)
Behavior Problems in School[4]	47	(68)	**	28	38

*$p < .05$
**$p < .01$
[1] Smoking 1/2 pack or more per day
[2] Drinking four or more drinks per week
[3] Use of marijuana more than twice per month, or other drug
[4] Truancy, sloughing, or fighting at school

SOCIAL RELATIONSHIPS

The fact that our subjects were in some senses troubled youths does not mean that they were unwilling to assume responsibility for their actions, however. Indeed, there was a striking willingness on the part of the young women not to seek elective abortions and on the part of their partners to assume some responsibility for their partners and children. As one would expect from national statistics, a very small number of these individuals (22, or 8%) were married at the time of conception. However, this does not mean that the others were involved in casual relationships. On the contrary, the vast majority described themselves as engaged, living together, or steadily dating, and a considerable portion (110, or 43%) of the 253 couples who were not married at the time of conception chose to legitimize the birth by getting married between conception and delivery. This "legitimation rate" is substantially higher than the national average, and thus mandates caution about generalizing from our sample to adolescent parents in markedly different ecologies.

In a recent set of analyses, Lamb, Elster, Peters, Kahn, and Tavare (1986) made the following series of comparisons:

initially married vs. not married at conception;
initially married vs. married between conception and delivery regardless of intensity of relationship before conception;
steady daters who chose to get married before delivery vs. daters who maintained a dating relationship.

Some of these data are presented in Tables 6.3 and 6.4. They show that individuals in each of these constellations have characteristics that alert one to and presage potential problems. Those who were married at the time of conception, for example, were much more likely to have left school prematurely and thus to have limited their long-term earning potential (see below), even though at this point they were significantly better off financially than those who were not married at conception. However, these two groups did not differ with respect to either parent's smoking, drinking, or substance abuse; history of physical or sexual abuse; arrest histories; and incidence of teen pregnancy or marital dissolution in their families of origin. Initially married couples responded more positively, and had parents (the prospective grandparents) who responded more positively than did the legitimizers or those who were not married at delivery (see Table 6.3). Compared with those steady daters who chose not to get married, it is clear that the daters who married constituted a group of individuals who were perhaps unusually responsible; they were much less likely to have been involved in illegal activities, much more likely to be in school and to be employed, and much more likely to be performing well in school, academically and behaviorally. The daters and legitimizing daters did not differ with respect to their own or their parents' responses; their smoking, drinking, substance abuse, or physical/sexual abuse histories; contraceptive usage; and marital status or history of teen pregnancy in their families of origin.

TABLE 6.3 Comparison Between Couples Who Were and Were Not Married at the Times of Conception and/or Delivery

	Married between conception & delivery (N=110)	Signifi- cance of difference	Married at conception (N=22)	Signifi- cance of difference	Not married at conception (N=253)
FOB's reaction to pregnancy (% positive)	47	+	75	**	38
MOB's reaction to pregnancy (% positive)	30	+	57		27
Reaction of FOB's parents (% positive)	36	**	85	**	34
Reaction of MOB's parents (% positive)	22	***	69	***	20
Change in relationship with FOB's parents (% better)	34		55	+	23
FOB's education status (% dropouts)	44	*	73	+	49
FOB's GPA (mean)	3.44	***	1.67	**	3.28
MOB's educational status (% dropouts)	45	**	77	**	42
MOB's GPA (mean)	3.31		3.33		3.13
FOB's salary/hour (mean)	4.88	*	7.97	***	4.56
MOB's hours worked/week (workers only) (mean)	19.0	+	26.67		22.67
FOB's present as support person during labor (%)	84		83	*	55
FOB's sexarche (mean in years)	15.94	*	12.67	*	15.64
MOB's sexarche (mean in years)	16	*	15.0	*	15.33
MOB's use of contraceptives at time of conception (% yes)	7		19	*	5

*** $p < .001$ by t-test or X^2 test; ** $p < .01$; * $p < .05$; + $p < .10$

TABLE 6.4 Comparison Between Regularly Dating Couples Who Chose to Marry or Not to Marry by Delivery

	Daters at Both Points (N = 29)	Signifi- cance of Differ- ence	Daters who Marry by Delivery (N = 84)
FOB's behavior problems in school (% yes)	87	**	36
FOB's hours worked/week (whole sample)	11.0	**	26.04
FOB present as support person during labor (%)	45	***	85
MOB identifies FOB as source of emotional support (%)	25	+	51
MOB's sexarche (mean age)	15.4	*	16.25
FOB's arrest history (% yes)	90	***	37

*** $p < .001$ by t-test or x^2-test
 ** $p < .01$
  * $p < .05$
  + $p < .10$

In all three groups, however, the partners' relationships appeared remarkably shaky. Each of the individuals was asked to name two people to whom he or she could turn for emotional support. Significantly, only 30 percent of the married fathers-to-be, 46 percent of the unmarried men, 52 percent of the legitimizers, 58 percent of the daters who married, and 53 percent of the steady daters identified their female partners; corresponding figures for the women were 29, 34, 54, 51, and 25 percent. These data suggest that these individuals—especially the adolescent mothers—still turn for emotional support to their families of origin rather than to their partners, whether or not they are married. Such findings surely raise questions about the prognosis of these relationships. Somewhat counterintuitively, the initially married couples were least likely to identify one another as sources of emotional support, whereas the daters who married were most likely to do so. This contradicts the hypothesis that marriages are healthier when they antedate and perhaps potentiate pregnancy than when they are precipitated by pregnancy. Nevertheless, few of these relatively young marriages appear destined to withstand the test of time.

PROGNOSIS OF RELATIONSHIPS

The doubts raised by these data are well borne out by prior research, showing quite clearly that marriages entered into in adolescence are much more likely to end in divorce than are marriages entered into by adult parents (Burchinal, 1965; Furstenberg, 1976). Recent analyses conducted by my colleagues and me using national data illustrate these effects. Teti, Lamb, and Elster (1987) showed that males whose first marriages occurred during their adolescence were disproportionately likely to have experienced a divorce, to have remarried, and to have had subsequent divorces than were men whose first marriages occurred after age 19. Identical effects were evident in cohorts of men who were in their 30s, 40s, and 50s at the time that these census survey data were collected in 1980. This suggests that the fragility of adolescent marriages is indeed a reliable phenomenon.

The data presented in the previous section (see Tables 6.3 and 6.4) raise interesting questions about which relationships are at greatest risk. Although one might have thought that those whose marriages were precipitated by premarital conception would be at greatest risk, the data available thus far do not support this hypothesis. Indeed, the legitimizing daters (see Table 6.4) appear to have more solid relationships and better prospects for financial security in the future than those who were married before the pregnancy occurred. In fact, it is not clear from these data why the latter group married, since their relationships do not seem at all strong.

EDUCATIONAL AND OCCUPATIONAL PROGNOSIS

One of the reasons why marriages formed in adolescence appear to be at great risk of dissolution is that the partners have to cope with somewhat shaky marriages in rather harsh economic and financial circumstances. Individuals who get married in adolescence suffer an educational and vocational disadvantage that remains with them throughout their lives (Card & Wise, 1978; Freedman & Thornton, 1979; Marsiglio, 1986; Teti et al., 1987). The magnitude of these effects on both mothers and fathers is illustrated in recent analyses by

Marsiglio (1986) of data from a nationally representative panel survey of 6000 males. They show that boys who were involved in adolescent pregnancies were much more likely to drop out of school than were their peers who fathered children in their 20s or had not fathered children by the end of the study. Interestingly, the probabilities of dropping out were very similar for married and unmarried fathers, and did not vary depending on whether the fathers lived with their young children. Presumably, these disadvantages have an effect on the present and future earning capacities of these young fathers.

In another set of analyses, Teti and colleagues (1987) used census data to show that these educational and vocational disadvantages persist well into adulthood. They showed that mature and middle-aged men were likely to be earning less and to have achieved less formal education when they had been involved in adolescent marriages than when they had not been so involved. The same effects were found in all the age cohorts studied, which include men now in their early 30s, now in their early 40s, and now in their early 50s. The legacy of an adolescent pregnancy lasted as long as 30 to 40 years—essentially an occupational lifetime.

These analyses leave the direction of effects unclear: *Does adolescent pregnancy force young men and women to quit schooling in order to care for their children and support their families, or do individuals who are disillusioned with education become involved in adolescent pregnancies?* Typically, the former interpretation has been offered, and Marsiglio (1986) presented regression analyses providing some support for this interpretation. Some of the data that we have obtained from the parents in the Teen Mother and Child Program suggest that the reverse may be true, however. The data presented in Table 6.1 show that an unusually high proportion of the individuals who were involved in adolescent pregnancies had a history of behavior problems at school and of contacts with the police, and had dropped out of high school by early in the pregnancy. These data suggest that disillusionment with education may lead some individuals into a life-style in which they risk teen-age pregnancy, rather than that teen pregnancy precipitates the termination of formal schooling. Indeed, there may be two discrete groups with very different characteristics, needs, and prognoses—one in which pregnancy precipitates school-leaving and one in which school-leaving precedes pregnancy. Unfortunately, however, none of the available data are really suitable for testing different causal models, so the interpretations must remain speculative for the present. The interpretations have crucial implications, not least of all for the design of sensitive and successful interventions.

Whatever their interpretations, however, findings such as these highlight the difficulties inherent in characterizing any factor as individual, micro, meso, or macro. Educational attainment, for example, is an individual variable, but it not only reflects the socioeconomic circumstances of the family of origin but also itself affects the future earning potential, residential location, and web of network relationships of the individuals and their families (Cochran, this volume). This suggests that it is less important to specify the level to which a characteristic refers than it is to consider a variety of potentially important aspects of individuals and their social ecologies.

IMPLICATIONS FOR PARENTING
AND CHILD DEVELOPMENT

In the preceding sections, we have described several situational and psychological factors that are likely to have an adverse impact on the ability of young parents to raise their children successfully. In this section, we discuss the association between these factors and adolescent parenthood, showing how they may affect parental sensitivity and effectiveness.

Stress and Coping

There is both conceptual and empirical justification for believing that maternal stress adversely affects parental sensitivity and thus the quality of infant–parent relationships (Lamb & Easterbrooks, 1981; Lamb & Gilbride, 1985; Ragozin, Basham, Crnic, Greenberg, & Robinson, 1982; Thompson, Lamb, & Estes, 1982; Vaughn, Egeland, Sroufe, & Waters, 1979). Belsky (1980), Crockenberg (1981), Goldberg and Easterbrooks (1984), and Garbarino (1977) have reported relationships between levels of stress and quality of parenting in adult-age parents. Vaughn and colleagues (1979) and Thompson and associates (1982) have independently demonstrated, in adult samples, that stress and major changes in family circumstances lead to changes in the security of infant–mother attachment. Not surprisingly, therefore, Ragozin and associates (1982) and Crnic, Greenberg, and Ragozin (1981) found a significant relationship between reported stress at one month postpartum and maternal sensitivity to infant cues three months later. When other variables such as maternal age and social support were included in the analyses, however, the effect of stress was greatly minimized. Likewise, Elster, Montemayor, and Gilbride (1984) found that degree of stress accounted for only a small portion of the variance in maternal sensitivity during a feeding, once variations in age and SES were statistically controlled. Further, Lamb and Elster (1985) found no relationship between measures of mother–infant engagement and responsiveness, on the one hand, and measures of marital quality, life stress, and social support, on the other. This may well be because Lamb and Elster's observations were made in a triadic context, whereas all other studies involved dyadic observations.

Other findings indeed suggest that the correlates of parental behavior in dyadic and triadic contexts are different. Belsky, for example, found that whereas the quality of maternal behavior in a dyadic context predicted later infant behavior in Ainsworth's Strange Situation procedure (Belsky, Rovine, & Taylor, 1984), the quality of maternal behavior in a triadic context did not (Belsky, personal communication). If this finding is replicated in other studies, it would underscore the need for multisituational assessments of parent–infant interaction in studies designed to explore the effects of stress, social support, and the like. It would also point to the need for care in the generalization and interpretation of findings obtained in studies sampling a narrow range of situations.

In our study of adolescent mother–father–infant triads, furthermore, only the measures of paternal engagement and involvement were related to measures of social support and social stress (Lamb & Elster, 1985). Thus the quality of the social context (e.g., stress and social support) appeared to have much less influence on maternal behavior than on paternal behavior. Belsky and colleagues (1984) reported similar patterns of correlations in their research on adult parents.

For various reasons, adolescent parents may be faced with stresses with which they are not equipped to cope. First, the timing of the pregnancy is out-of-phase with the usual course of life events. Excessive stress frequently results when role transitions, such as movement into parenthood or marriage, occur sooner or later than the socially prescribed norms (Bacon, 1974; Russell, 1980). Although teen-age pregnancy is now more accepted than in previous years (see above), it is still considered a "problem" and is not completely sanctioned by the adolescents' family or community. This is especially true for those teens who have not yet graduated from high school. In addition, adolescent parents, particularly younger ones, are faced with a variety of situational crises (pregnancy, parenthood, and marriage), which are superimposed on a maturational "crisis" (adolescence), each of which may be associated with some degree of stress (Dyer, 1965; Frank & Cohen, 1979; Judd, 1967; Miller & Sollie, 1980; Parad & Caplan, 1965; Teti & Lamb, 1986). These developmental stresses combine to affect adolescents who may have neither the psychological maturity nor the social support to help them cope adequately.

The vocational-educational disadvantage suffered by many adolescent parents exaggerates the stresses they face. Compounding these financial problems is the fact that adolescent parents, especially those who conceive premaritally, have more rapid subsequent pregnancies and ultimately have larger families than parents who do not conceive premaritally or who postpone childbearing (Moore & Hofferth, 1978; National Council on Health Statistics, 1981). Thus women who begin childbearing at fifteen years of age or younger have one-and-a-half to three times as many children as women who have their first child in their early twenties (Moore & Hofferth, 1978). Financial concerns, large family sizes, and marital instability may all impose stresses on adolescent parents that adversely affect their parental effectiveness. Unfortunately, the situation may be exacerbated by the absence or insufficiency of social support.

Social Support

Although it has long been known that the availability of social support helps to maintain mental health, researchers have only recently started asking whether social support affects parental behavior (Cochran & Brassard, 1979; Henderson, Byrne, & Duncan-Jones, 1980; Hirsch, 1980). Among a group of adolescent mothers, Colletta and Gregg (1981) found a positive correlation between the total amount of social support and the frequency of appropriate maternal behavior. Of the various kinds of perceived supports, the most important was emotional support, especially when provided by the mother's family of origin. Crnic and colleagues (1981) found that the amount and quality of perceived support were directly related to mothers' responsiveness during interactions with their infants.

The role of family support in helping adolescent parents was also studied by Furstenberg and Crawford (1978) in their five-year longitudinal investigation of adolescent mothers. Major differences were found between never-married women who lived with parents or relatives and those who lived alone. The former were more likely to have returned to or have graduated from school, to be employed, and to be independent of welfare support. Furstenberg and Crawford concluded that there is a strong relationship for never-married young mothers between their residential situation and their vocational-educational outcomes. Apparently, more material support and help with child care are available to mothers who live at home. No relationship was found, however, between where the teen mothers lived and either their child-rearing patterns or the developmental status of their infants.

As mentioned earlier, however, strong associations between reported social support and quality of mothering were not found in two studies of the parental behavior of adolescent parents, even when an attempt was made to look at the combined impact of stress and social support (Elster et al., 1984; Lamb & Elster, 1985). In the second study, however, scores on some measures of social support were correlated with indices of paternal, but not maternal, behavior. One reason for those findings may be that the predominantly white subjects came from more advantaged backgrounds and had more advantaged circumstances, in terms of life stresses, social support, education, and occupation, than the black disadvantaged subjects studied by most researchers.

One aspect of the association between social support and adolescent parental behavior has to do with the degree of support provided by males to their adolescent partners. The lack of appreciation by health care providers, social service workers, and the mothers' parents for the problems experienced by teen fathers often results in their total exclusion from prenatal care and, in some cases, the denial of opportunities to see their female partners or children. Furstenberg's (1976) longitudinal study of adolescent mothers indicated that parenthood frequently led teens, regardless of marital status, to establish their own households, thus separating them from potentially useful social supports. Five years after delivery, 26 percent of the teen mothers (11% never married and 15% previously married) were living alone. This is not to imply, of course, that the children of teen-age parents are better off in all respects when their parents marry: As described earlier, couples who marry in adolescence are much more likely to separate than are couples who marry later (Burchinal, 1965; Moore, Waite, Hofferth, & Caldwell, 1978; Teti et al., 1987). This is a source of concern because poor relationships between parents seem to have a more predictably negative effect on child development than any other childhood experiences, including divorce and single parenthood (Rutter, 1973; Rutter, Cox, Tupling, Berger, & Yule, 1975).

Of course, financial, emotional, and social stress and social isolation are not the only factors that affect parental competence, although I believe that they are the most important in the context of deficiencies in the parental behavior of adolescent mothers. Other factors of special relevance to adolescents are the parents' degree of cognitive and socio-emotional maturity (Adams & Jones, 1981; Elkind, 1974; Sadler & Catrone, 1983; Stevens & Duffield, 1986), attitudes toward child rearing (Field, Widmayer, Stringer, & Ignatoff, 1980; Green, Sandler, Altemeier, & O'Connor, 1981; Mercer, Hackley, & Bostrom, 1984; Wise & Grossman, 1980; Zuckerman, Winesmore, & Alpert, 1979), and knowledge of child development (Chamberlain et al., 1979; Epstein, 1980; Field et al., 1980; Parks & Smeriglio, 1983). Characteristics of the infant—including temperament and gender—are also likely to affect the behavior of both adolescent and adult parents.

CONCLUSION

There is ample evidence that many teen-age parents confront an unsupportive ecology that threatens to disrupt their lives and relationships. By definition, adolescent mothers (and often their partners as well) are thrust into parental roles before they have resolved their "identity crises" and established their life goals. Although social rejection is not as widely encountered as in the past, social disapproval remains, and federal welfare policy financially punishes poor couples who marry and try to support one another emotionally. The relationships between the parents are often in their infancy, and regardless of the degree of formal intracouple commitment manifest by delivery, the parents are not likely to see one another as reliable sources of emotional support. Many adolescent parents are still in school when pregnancy occurs, and this poses a cruel dilemma: If they persist at school, they are limited with respect to both the number of hours they can work and the amount they are likely to be paid per hour, whereas if they drop out of school, they can earn more in the short term but face a lifelong limitation on their earning power. The financial stresses associated with these circumstances, imposed as they are on shaky relationships between immature partners, simply magnify the risks of marital dissolution. Thus the ecology is often hostile at all—macro, meso, micro, and individual—levels.

Nevertheless, the circumstances of adolescent parents are not necessarily or uniformly bad. As reported above, the adverse factors are often cumulative and may compound one another, but by the same token some positive or supportive circumstances can offset the adverse effects of those that are negative. Not all adolescent parents perform inadequately: Indeed, the majority provide adequate or good parental care. The quality of parental behavior—among adults as among adolescents—is multiply determined, and from the perspectives of both researchers and interventionists, it is important to appraise all factors of potential relevance before identifying the origins of deficiencies or the focus of intervention. One implication of the findings presented in this chapter is that successful interventions must be multifaceted and individually tailored in light of the particular circumstances. Conversely, the same intervention can have different effects on different families or on similar families in different circumstances.

REFERENCES

Adams, G. & Jones, R. M. (1981). Imaginary audience behavior: A validation study. *Journal of Early Adolescence, 1,* 1–10.

Bacon, L. (1974). Early motherhood, accelerated role transition, and social pathologies. *Social Forces, 52,* 333–41.

Bates, J. E.; Maslin, L. A.; & Frankel, K. A. (1985). Attachment security, mother–child interaction, and temperament as predictors of behavior problem ratings at age three years. In I. Bretherton & E. Waters (Eds.), *Growing points of attachment theory and research. Monographs of the Society for Research in Child Development, 50,* serial number 209.

Battaglia, F. C.; Frazier, T. M.; & Hellegers, A. E. (1963). Obstetric and pediatric complications of juvenile pregnancy. *Pediatrics, 32,* 902–10.

Belsky, J. (1980). Child maltreatment: An ecological integration. *American Psychologist, 35,* 320–35.

Belsky, J.; Rovine, M.; & Taylor, D. G. (1984). The Pennsylvania Infant and Family Development Project, III: The origins of individual differences in infant–mother attachment: Maternal and infant contributions. *Child Development, 55,* 718–28.

Burchinal, L. G. (1965). Trends and prospects for young marriages in the United States. *Journal of Marriage and the Family, 27,* 243–54.

Card, J. J. & Wise, L. L. (1978). Teenage mothers and teenage fathers: The impact of early childbearing on the parents' personal and professional lives. *Family Planning Perspectives, 10,* 199–205.

Chamberlain, R. W.; Szumoski, E. K.; & Zastowny, T. R. (1979). An evaluation of efforts to educate mothers about child development in pediatric
office practices. *American Journal of Public Health, 69,* 875–85.

Cochran, M. M. & Brassard, J. A. (1979). Child development and personal
social networks. *Child Development, 50,* 601–16.

Colletta, N. D. & Gregg, C. H. (1981). Adolescent mothers' vulnerability to
stress. *Journal of Nervous and Mental Disease, 169,* 50–54.

Crnic, K. A.; Greenberg, M. T.; & Ragozin, A. S. (1981). The effects of stress
and social support on maternal attitudes and the mother–infant relationship. Paper presented at the Biennial Meeting of the Society for Research in Child Development, Boston.

Crockenberg, S. B. (1981). Infant irritability, mother responsiveness, and
social support influences on the security of infant–mother attachment.
Child Development, 52, 857–65.

DeLissovoy, V. (1973). Child care by adolescent parents. *Child Today, 2,*
23–25.

Dyer, E. D. (1965). Parenthood as crisis: A re-study. In H. J. Parad (Ed.),
Crises intervention: Selected readings (pp. 312–23). New York: Family
Service Association of America.

Elkind, D. (1974). *Children and adolescents: Interpretive essays on Jean Piaget.*
New York: Oxford University Press.

Elster, A. B. & Lamb, M. E. (Eds.) (1986). *Adolescent Fatherhood.* Hillsdale,
NJ: Lawrence Erlbaum Associates.

Elster, A. B.; Montemayor, R.; & Gilbride, K. E. (1984). Factors influencing
the parental behavior of adolescent mothers. Unpublished manuscript,
Department of Pediatrics, University of Utah Medical Center.

Elster, A. B.; Lamb, M. E.; Peters, L.; Kahn, J.; & Tavare, J. (1987). Judicial
involvement and conduct problems of fathers of infants born to adolescent mothers. *Pediatrics, 79,* 230–34.

Epstein, A. S. (1980). *Assessing the child development information needed by
adolescent parents with very young children.* Final Report, U.S. Department of Health, Education and Welfare.

Field, T. M.; Widmayer, S. M.; Stringer, S.; & Ignatoff, E. (1980). Teenage,
lower-class, black mothers and their preterm infants: An intervention
and developmental follow-up. *Child Development, 51,* 426–36.

Frank, R. A. & Cohen, D. J. (1979). Psychosocial concomitants of biological
maturation in preadolescence. *American Journal of Psychiatry, 136,*
1518–24.

Freedman, D. S. & Thornton, A. (1979). The long-term impact of pregnancy
at marriage on the family's economic circumstances. *Family Planning
Perspectives, 11,* 6–21.

Furstenberg, F. F. (1976). The social consequences of teenage parenthood.
Family Planning Perspectives, 8, 148–64.

Furstenberg, F. F. & Crawford, A. G. (1978). Family support: Helping teen-
age mothers to cope. *Family Planning Perspectives, 10,* 322–33.

Garbarino, J. (1977). The human ecology of child maltreatment. *Journal of Marriage and the Family, 39,* 721–36.

Goldberg, W. A. & Easterbrooks, M. A. (1984). Role of marital quality in toddler development. *Developmental Psychology, 20,* 504–14.

Green, J. W.; Sandler, H. M.; Altemeier, W. A.; & O'Connor, S. M. (1981). Child rearing attitudes, observed behavior, and perception of infant temperament in adolescent versus older mothers. *Pediatric Research, 15,* 442.

Gutelius, M. F.; Kirsch, A. D.; Macdonald, S.; Brooks, M. R.; & McErlean, T. (1977). Controlled study of child health supervision: Behavioral results. *Pediatrics, 60,* 294–304.

Henderson, S.; Bryne, D. G.; & Duncan-Jones, P. (1980). Social relationships, adversity, and neurosis: A study of associations in a general population sample. *British Journal of Psychiatry, 136,* 574–83.

Hirsch, B. J. (1980). Natural support systems and coping with major life changes. *American Journal of Community Psychology, 8,* 159–72.

Israel, S. L. & Woutersz, T. B. (1963). Teenage obstetrics: A comparative study. *American Journal of Obstetrics and Gynecology, 83,* 659–68.

Jessor, S. L. & Jessor, R. (1975). Transition from virginity to nonvirginity among youth: A social-psychological study over time. *Developmental Psychology, 11,* 478–84.

Jones, E. F.; Forrest, J. D.; Goldman, N.; Henshaw, S. K.; Lincoln, R.; Rosoff, J. I.; Westoff, C. F.; & Wulf, D. (1985). Teenage pregnancy in developed countries: Determinants and policy implications. *Family Planning Perspectives, 17,* 53–63.

Judd, L. L. (1967). The normal psychological development of the American adolescent. *California Medicine, 107,* 465–70.

Lamb, M. E. & Easterbrooks, M. A. (1981). Individual differences in parental sensitivity: Origins, components, and consequences. In M. E. Lamb & L. R. Sherrod (Eds.), *Infant social cognition: Empirical and theoretical considerations.* Hillsdale, N.J.: Lawrence Erlbaum Associates.

Lamb, M. E. & Elster, A. B. (1985). Adolescent mother–infant–father relationships. *Developmental Psychology, 21,* 768–73.

Lamb, M. E.; Elster, A. B.; Peters, L. J.; Kahn, J. S.; & Tavare, J. (1986). Characteristics of married and unmarried adolescent mothers and their partners. *Journal of Youth and Adolescence, 15,* 487–96.

Lamb, M. E. & Gilbride, K. (1985). Compatibility in parent–infant relationships: Origins and processes. In W. Ickes (Ed.), *Compatible and incompatible relationships* (pp. 33–60). New York: Springer.

Levine, L.; Garcia Coll, C. T.; & Oh, W. (1985). Determinants of mother–infant interaction in adolescent mothers. *Pediatrics, 75,* 23–29.

Marsiglio, W. (1986). Teenage fatherhood: High school accreditation and educational attainment. In A. B. Elster & M. E. Lamb (Eds.), *Adolescent Fatherhood* (pp. 67–87). Hillsdale, N.J.: Lawrence Erlbaum Associates.

Mercer, R. T.; Hackley, K. C.; & Bostrom, A. (1984). Adolescent motherhood: Comparison of outcome with older mothers. *Journal of Adolescent Health Care, 5,* 7–13.

Miller, B. C. & Sollie, D. L. (1980). Normal stresses during the transition to parenthood. *Family Relations, 29,* 459–65.

Moore, K. A. & Hofferth, S. L. (1978). *The consequences of age at first childbirth: Family size.* Washington, D.C.: The Urban Institute.

Moore, K. A.; Waite, L. J.; Hofferth, S. L.; & Caldwell, B. M. (1978). *The consequence of age at first childbirth: Marriage, separation, and divorce.* Washington, D.C., The Urban Institute.

National Center for Health Statistics (1981). Socioeconomic differentials and trends in the timing of births. *Vital Health Statistics 23*(6).

National Center for Health Statistics (1984). *Monthly vital statistics report. Advanced report of final natality statistics, 1982.* Washington, D.C.: U.S. Government Printing Office, *33*(6) (Supplement).

Nakashima, I. I. & Camp, B. W. (1984). Fathers of infants born to adolescent mothers: A study of paternal characteristics. *American Journal of Diseases of Children, 138,* 452–54.

Parad, H. J. & Caplan, G. (1965). A framework for studying families in crises. In H. J. Parad (Ed.), *Crises intervention: Selected readings* (pp. 53–74). New York: Family Service Association of America.

Parks, P. L. & Smeriglio, V. L. (1983). Parenting knowledge among adolescent mothers. *Journal of Adolescent Health Care, 4,* 163–67.

Pauker, J. D. (1971). Fathers of children conceived out of wedlock: Prepregnancy, high school, psychological test results. *Developmental Psychology, 4,* 215–18.

Ragozin, A. S.; Basham, R. B.; Crnic, K. A.; Greenberg, M. T.; & Robinson, N. M. (1982). Effects of maternal age on parenting role. *Developmental Psychology, 18,* 627–34.

Rivara, F. P.; Sweeney, P. J.; & Henderson, B. F. (1985). A study of low socioeconomic status, black teenaged fathers and their non-father peers. *Pediatrics, 75,* 648–56.

Russell, C. S. (1980). Unscheduled parenthood: Transition to 'parent' for the teenager. *Journal of Social Issues, 36,* 45–63.

Russell, G. (1978). The father role and its relation to masculinity, femininity, and androgyny. *Child Development, 49,* 755–65.

Rutter, M. (1973). Why are London children so disturbed? *Proceedings of the Royal Society of Medicine, 66,* 1221–25.

Rutter, M.; Cox, A.; Tupling, C.; Berger, M.; & Yule, W. (1975). Attainment and adjustment in two geographic area, I: The prevalence of psychiatric disorder. *British Journal of Psychiatry, 126,* 493–509.

Sadler, L. S. & Catrone, C. (1983). The adolescent parent: A dual developmental crisis. *Journal of Adolescent Health Care, 4,* 100–05.

Stevens, J. H. (1985). *Parenting skill: Does social support matter?* Paper presented to the Society for Research in Child Development, Toronto.

Stevens, J. H. & Duffield, B. N. (1986). Age and parenting skill among poverty black women. *Early Childhood Research Quarterly,* in press.

Teti, D. M., & Lamb, M. E. (1986). Sex role development in adolescent males. In A. B. Elster & M. E. Lamb (Eds.), *Adolescent fatherhood*. Hillsdale, NJ: Erlbaum.

Teti, D. M.; Lamb, M. E.; & Elster, A. B. (1987). Long-range socioeconomic and marital consequences of adolescent marriage in three cohorts of adult males. *Journal of Marriage and the Family, 49*, 499–506.

Thompson, R. A.; Lamb, M. E.; & Estes, D. (1982). Stability of infant–mother attachment and its relationship to changing life circumstances in an unselected middle class sample. *Child Development, 53*, 144–48.

Vaughn, B.; Egeland, B.; Sroufe, L. A.; & Waters, E. (1979). Individual differences in infant–mother attachment at 12 and 18 months: Stability and change in families under stress. *Child Development, 50*, 971–75.

Vinovskis, M. A. (1986). Adolescent sexuality, pregnancy, and childbearing in early America: Some preliminary speculations. In J. B. Lancaster & B. A. Hamburg (Eds.), *School-age pregnancy and parenthood: Biosocial dimension*. Chicago: Aldine.

Wise, S. & Grossman, F. K. (1980). Adolescent mothers and their infants: Psychological factors in early attachment and interaction. *American Journal of Othopsychiatry, 50*, 454–68.

Zuckerman, B.; Winesmore, G.; & Alpert, J. J. (1979). A study of attitudes and support systems of inner city adolescent mothers. *Journal of Pediatrics, 95*, 122–25.

7

CLIENT CHARACTERISTICS
AND THE DESIGN OF
COMMUNITY–BASED INTERVENTION PROGRAMS

DOUGLAS R. POWELL

Defining "program" as the unit of analysis, Douglas Powell examines the interaction between a parent-education and support-intervention program and both its actual and intended participants. Powell argues that program processes that are insensitive to and not interactive with participant involvement processes can weaken the magnitude of intervention effects. Through the experiences of the Child and Family Neighborhood Program, he notes the challenges that can beset an ecologically sensitive intervention.

This chapter offers an ecological perspective on the interactions between community-based intervention programs and their client* environments. A premise of the chapter is that advances in the design of voluntary parent–child intervention programs for high-risk communities depend partly on a systematic understanding of how programs are organized to generate and serve participants. In this chapter, the unit of analysis is a developing program. Theories about organization–environment relations provide a point of departure for examining program perceptions of and responses to the characteristics of intended and actual participants.

The relationship of program design to participant characteristics is an increasingly important topic in the early intervention field.

---

*Use of the term "client" in this chapter is not to be associated with the negative connotation sometimes attributed to the word. The words "client" and "participant" are used interchangeably. The intervention and research described in this chapter were supported by a grant from the W. K. Kellogg Foundation.

Since the 1970s there has been a trend toward child and family programs that respond to community needs and values rather than impose a monolithic treatment on individuals in diverse communities (Schaefer, 1977). Program sensitivity to the cultural norms of ethnic and low-income populations (Laosa, 1984; Rogler, Malgady, Costantino, & Blumenthal, 1987), and movement toward co-equal relations between program workers and participants (Cochran, this volume; Powell, 1984) are among the developments that heighten interest in how the organization of an intervention program facilitates or impedes responsiveness to participant needs and characteristics. Findings of recent studies of program processes suggest that program implementation problems have hampered the provision of services and perhaps weakened the magnitude of intervention effects (Mindick, 1986; Travers, Irwin, & Nauta, 1981). It appears that the ways in which programs recruit participants and provide services are as important to investigate as program effects on individuals.

The ecological paradigm offers a conceptual tool for thinking about the response of intervention programs to their participants. The conventional model of program development and research is ill-suited to address questions about interactions between programs and participants. The dominant program design and evaluation paradigm sets forth an image of the program as a static entity that acts upon compliant individuals who enter the program as blank slates. An ecological approach to program design and research challenges this view of programs and participants. It calls for a conception of program participants as active individuals who engage a program in different ways, and an image of the program as a dynamic, fluid organization.

A key to the development of community-based programs is responsiveness to the needs and characteristics of intended and actual participants. It is assumed the program will develop a structure and content to attract and sustain the involvement of all eligible clients. Participant recruitment and retention experiences are a crucial test of a program's design. What are the processes through which a program accommodates the desired participants? What types of recruitment strategies reach what types of individuals? It is rare for reports of program research to discuss participant attrition or the characteristics of individuals not attracted to a given program.

This chapter describes and analyzes the experiences of the Child and Family Neighborhood Program (Powell, 1987) as a retrospective case study of the interplay between client characteristics and the design of a community-based parent education and support program. The intent is to identify program design issues and research directions regarding the recruitment and participation of individuals in a community-based intervention. The chapter, then, pertains to one element (client pool) of the program's environment. It is beyond the scope of this chapter to examine all aspects of program-environment relations (for example, linkages with funding agencies) that affect program design and operations.

The literature on organization behavior provides a useful framework for analyzing relations between participants and community-based intervention programs. In the 1960s the theoretical approach to the study of organizations shifted to an open systems view (Katz & Kahn, 1966; Thompson, 1967), giving priority to the interrelationships of organizations and their environments. Organizational viability came to be viewed as largely a function of the relationships maintained with the environment. The major environmental element of concern to this chapter is the population to be served. Program success in attracting and sustaining a sizable group of participants is a form of environmental sanctioning of the intervention. Program survival and legitimacy are at stake. Measures of enrollment, attendance, and attrition do not constitute indicators of program effects, but they do represent environmental support of program functions. As experience with intervention programs has long suggested, securing clients in high-risk communities is no easy task (Chilman, 1973).

A critical question is what type of internal arrangement of staff roles, communication patterns, and decision-making structure enables an organization to learn about and respond to characteristics of its client pool. That is, what mechanisms permit an organization to cope with uncertainty and change in its social context? Organizational learning must occur rapidly and accurately in the fledgling period of operations (Argyris & Schon, 1978; Miles & Randolph, 1980). A prevalent view in the organizational literature is that in new or turbulent environments, informal and flexible systems of internal communication and change are likely to foster adaptations of the organization to its environment. For example, a classic study of organization–environment relations suggested that an organic model of organization—where there is increased lateral communication, less clearly defined roles, and more decision making down the line—is appropriate in environments characterized by novelty or change (Burns & Stalker, 1961). More recently, however, questions have been raised about the utility of egalitarian staff relations for social intervention programs that have a short life-span and lack a well-trained, homogeneous staff from the outset (Mindick, 1986). Minimalist hierarchical arrangements can deter prompt decision making and inhibit needed supervision of staff.

The Child and Family Neighborhood Program (CFNP) utilized a typical strategy for developing a community-based program. It began with an operational plan that was based on perceptions of the needs and characteristics of desired participants, and theory, logic, and intuition regarding intervention methods. The original program design was to undergo modification if it did not prove to be responsive to the interests of actual and anticipated participants.

The first part of this chapter describes the design and implementation of the CFNP, with particular focus on the role of staff perceptions of participant characteristics. The chapter explores how assumptions about the needs and characteristics of potential users influenced the shape of the original program design. It also examines the ways in which the program responded to the realities of the population group to be served. A critique of the appropriateness of the original program design in light of implementation experiences is included. The last part of the chapter sets forth implications for the design and evaluation of community-based programs, with specific attention given to measurement issues.

The data base of this chapter is staff records and a series of detailed interviews with staff, supplemented by quantitative analyses of participant behaviors and characteristics. The views and interpretations of program experiences presented are those of the author, who was the project's co-founder and first director.

PROGRAM DESIGN AND IMPLEMENTATION

Design Decisions

The Child and Family Neighborhood Program (CFNP)was based in a compact, white, suburban Detroit neighborhood initially identified by school officials as having a large number of children with low academic achievement and of socially isolated low-income families. Early and long-term intervention aimed at mothers of very young children was viewed as a way to prevent later school problems. The program sought to serve all mothers in the neighborhood with an infant six months or younger. It was developed by The Merrill-Palmer Institute in cooperation with the Wayne-Westland

Community Schools. Representatives of both institutions contributed to the development of the program design. This chapter focusses on program design decisions that related most directly to perceptions of participant attributes. A detailed discussion of the intervention design is available elsewhere (see Powell, 1987; in press).

From its beginning, the CFNP adhered strongly to the value of individual change through peer group interaction. Consistent with much of the literature on adult education (Brookfield, 1986), group discussion was viewed as an indispensable method of working with parents. Conceptually, this social-psychological approach reflected Lewin's (1958) theory regarding the influence of the peer group on individual behavior and change. It was endorsed as a needed strategy by public school officials who contributed to the design and implementation of the intervention via the school system's co-sponsorship of the program. School officials maintained a perception of community residents as geographically separated from their families and socially distant from their neighbors. The program's peer group was seen as a way to provide informal social support to parents. Frequent (twice weekly) group meetings for substantial amounts of time (two hours) were viewed as important to making the program an integral part of parents' lives.

During the time the program was being conceptualized, the program staff consulted with elected public officials and professionals serving the community. These individuals portrayed neighborhood residents as uniformly low income and dependent on welfare or unsteady employment; lacking in personal planning and follow-through skills; functioning in a turbulent home life, including physical violence and alcoholism; and involved in unstable marriages or cohabitation arrangements. In general, children were seen as unplanned, unwanted, and minimally cared for. A high-level school official suggested that the neighborhood was "just as bad as Harlem." Exemplary of these characterizations was the public comment of a City Council member when approval was sought to locate the project in a residential section of the neighborhood: "I wish you luck in fixing up the moral standards of the neighborhood."

While intervention method (group-based) and staff perceptions of prospective clients (socially isolated) converged to support the group discussion format, these two sources of guidance for the program design prompted major staff disagreements about the length and content of group meetings. The aforementioned images of the program's potential clients prompted some staff involved in designing the program to argue for advertising short-term (four- to six-week) classes on specific child and family topics. The rationale for offering short-term participation was that the mothers would be more willing to join a brief class than a group that involved a long-term commitment. It was argued that the transient character of the mothers' lives would preempt their consideration of sustained involvement. There also were suggestions that some of the mothers would not possess the needed ability for long-term personal involvement in a group; the idea would be personally threatening. The advertised short-term nature of the classes was seen as a "hook." Each class was to be followed by another class, so in essence participants would be involved over the long term if they continued to attend subsequent classes. The opposing argument was that short-term classes might take on a didactic, teacher-directed character that would violate principles and benefits of peer group discussion. Proponents of that argument thought that short-term sessions would negate the group stability and interpersonal relationships that a long-term arrangement might foster.

Disagreement regarding the content of group meetings focussed on the desirability of imposing topics versus having group members determine the topics to be discussed. The rationale for predetermined content reflected one or more of the following expectations: (a) Mothers might not be able to decide what they wanted to discuss; (b) some of the mothers would not know what they needed to know; and (c) not publicizing a specific topic would make the program too vague and unattractive. The alternative view was that member control of group life was expected to lead to group discussion of topics of genuine interest to participants and to contribute to higher rates of participant retention in the program as well as greater individual change. Also, it was thought that the imposition of topics might place mothers in student roles, thereby alienating those who had negative feelings toward school. Furthermore, staff determination of group content might move staff toward authoritarian roles, which could prove to be detrimental in maintaining program sensitivity to neighborhood values.

Staff deliberations led to a decision to implement long-term discussion groups without predetermined content. If this approach failed to attract sufficient members, then short-term sessions would be offered on specific topics.

New groups were to be led by paraprofessionals from the community who received training in group work and child development. The use of paraprofessionals indigenous to the community was viewed as a way to strengthen ties to the local neighborhood and to facilitate communication with prospective and actual participants. The co-sponsoring school district had effectively used paraprofessionals in a number of educational outreach programs, prompting school officials to argue persuasively for a program design that called for paraprofessionals to be the primary front-line workers in the intervention. In addition to the paraprofessionals, the staff was to include a community outreach social worker, a public health nurse, and a specialist in early childhood, all of whom would be available to participants for individual consultation.

Recruitment Processes

The initial months of program implementation revealed a more differentiated set of views of potential users than the uniform portrayals offered by community leaders initially. The first three months of program implementation included the gathering of ethnographic data on neighborhood characteristics, resources, and civic groups. The information was to help staff engage in the fine tuning of program implementation (e.g., determining the strengths and limitations of existing health and social services for parents of very young children in the neighborhood). The ethnographer took a strong grass roots approach, walking about the neighborhood daily to converse informally with people on the street. Neither the field reports nor subsequent participant recruitment efforts supported a blanket application of the generalized, if not stereotypic, characterizations of residents.

Staff came to think of three loosely defined types of families with young children living in the neighborhood. One type was extremely high risk in terms of environmental stress. Families faced multiple problems and approximated the stereotypic characterizations offered by community leaders. A second type could be described as working poor. The parents lived under difficult circumstances but had enough resources to provide a relatively stable home environment. The third type was solid working class, with the main financial provider employed in semi-skilled or skilled work. The staff's three-tiered perception of neighborhood families came about largely through efforts to recruit parents, not from information provided by civic leaders and professionals responsible for programs serving the neighborhood.

The initial weeks of program recruitment showed a strong relationship between type of recruitment technique and type of parent agreeing to participate. Brochures about the program distributed in the neighborhood attracted the interest of socially skilled, assertive mothers who were actively looking for some type of activity (not necessarily a parent program) to pursue outside the home. In these instances it was a matter of parents finding the program rather than the program finding parents. This passive recruitment strategy, however, did not reach the vast majority of mothers who joined the program. It became necessary to actively seek out the names of prospective participants and to engage in one-to-one contact with potential users, generally in their homes. Names of potential participants were secured from hospital records, schools, and other community agencies. Each individual was sent a letter describing the program, followed by a telephone call or, where phones were not available, a brief visit to the home to schedule a visit to discuss the program. These one-to-one recruitment efforts were carried out by the program's nurse and outreach social worker.

Slightly more than one-third of the mothers who joined the program received staff consultation services on an individual basis prior to participating in a group. In these cases the staff member making the initial recruitment contact deemed the mother in need of assistance surrounding medical or social service issues. Generally, staff help was provided in the context of a home visit. The rationale here was twofold: (a) The provision of services was seen as a mechanism for stabilizing stressful life circumstances that would inhibit participation in the group, and (b) it was hoped the mother would develop trust in the program via the staff member.

Figure 7.1 shows movement from a pool of prospective participants to the number who actually joined during a two-and-one-half-year period beginning with the implementation of the program. The pool of potential participants (n = 333) consisted of names of mothers generated through hospital records (56%), parents in the program (13%), schools (12%), agencies (4%), self-referral (3%), and other informal means. Sixty-eight percent of these individuals were contacted about the program in person or by telephone. Those not contacted reportedly had moved from the neighborhood (e.g., dwelling was empty or occupied by new party when staff visited). About 23 percent of those contacted were ineligible for participation due to the baby's age (older than six months). Almost one-half (47%) of the contacted eligibles did not join the program. Staff were not given a reason by 33 percent of the nonjoining mothers. Ten others acted in ways the staff found to be hostile. Others declined to participate due to employment or school (24), pressing life stress factors (7), busy schedules (9), or religious beliefs (1).

FIGURE 7.1   Participant Recruitment Experiences of CFNP (from October 15, 1978 to January 30, 1981)

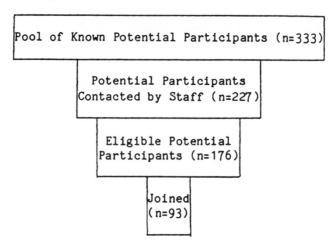

The factors that prospective participants considered in deciding whether to join the program were generally unknown to the staff. It appears that parents' social networks came into play in at least several ways. There were instances where mothers routinely and frequently congregated with family and friends. One individual, for example, socialized with her own mother and sisters several times a week. The program, she reported, would be duplicative of these gatherings. It also seemed that significant members of one's social network had an influence on the decision about joining. It was not uncommon to learn that a person's husband or mother did not approve of the program. There also were cases of a mother's relatives (especially her own mother) encouraging program participation but the mother not wanting to join.

Only about one-half of those mothers who received staff consultation following an initial recruitment contact actually joined a group. One such nonjoiner, for instance, received a total of 15 contacts with staff through home visits (9) and telephone calls (6) surrounding the baby's health (for instance, concern about ingrown toenails). During the fourteenth contact (a home visit), a staff member told the mother of a forthcoming group she could join. One of the staff members present at this visit later recalled that the staff comment prompted a response from the mother that resembled "a light being turned off." The mother's nonverbal behaviors reportedly showed signs of immediate disengagement from the interaction, and she would not discuss the group with the staff members.

In terms of the three-tiered view of neighborhood residents that was eventually held by the staff, the program attracted and sustained the participation of working poor mothers to a considerably greater extent than working-class mothers or mothers from multi-problem households. The program was not able to secure the participation of the vast majority of mothers from multi-problem families who were referred to it. The transient nature of this population group worked against program participation. Mobility was not the only deterrent, however. Some life-styles did not easily coincide with the structure imposed by the program (i.e. twice-weekly meetings at a program site). On one recruitment visit, for example, staff members arrived to find four adults and two babies in a one-room apartment with little furniture and a good deal of garbage and flies. The prospective participant did not make eye contact with the staff, appeared not to be listening, and yet encouraged the staff members to remain when they asked whether the time was good for a visit. There was one follow-up contact with the parent, who indicated no interest in the program or staff help. With another case, the staff had contact with a mother over a three-month period. During this time she moved out of her mother's house to live with a boyfriend, her father died, her boyfriend lost his job, her only car stopped running, the baby was hospitalized for an illness, and she experienced difficulty getting up before noon. She and her boyfriend visited group meetings on two separate occasions. A staff report indicated that they sat "stiff in their chairs, never moved, wouldn't speak to others but yelled at her cranky baby."

At the other end of perceived family types in the neighborhood (solid working-class), the program also experienced some problems in securing participants. Here there was a tendency for prospective participants to suggest, "Your program sounds great. The people who need you live over there." Apparently in these instances the program was viewed as a remedial project for the "have-nots." There were cases of the "have" mothers visiting the program once or twice, resulting in a decision not to join. For instance, one day a working-class mother visited an ongoing group to "see what it's all about." She was stunned to see one of the mothers in the group use a rag for her child's diaper, and firmly told a program worker, "I'm too good for this." She never returned.

Participant Experiences

How well did the intervention design fit the characteristics of parents who joined the program? It is questionable whether the early months of discussion group participation were useful for mothers experiencing acute environmental stress, such as marital problems. An examination of variations in program participation found that high-stress mothers had lower levels of attendance and verbal participation at group meetings than mothers experiencing low levels of environmental stress. The former also consulted extensively with staff on an individual basis. These consultations focussed largely on medical and social service needs, and frequently took place during home visits.

A particularly noteworthy finding was that high-stress mothers exhibited a delayed integration into group life that followed by about six months the modal pattern of low-stress mothers. The integration process for low-stress mothers involved their reporting experiences with child rearing and parenthood to fellow group members in the first three or four months of group participation. This verbal behavior (which we labeled narrative behavior) was correlated with the subsequent formation of interpersonal ties with program peers at six months. For the high-stress mothers, the reporting of child-rearing and parenting experiences to group members occurred about six months after joining the program and was correlated with the formation of peer ties at about 12 months (Eisenstadt & Powell, 1987). The delayed integration of the high-stress mothers into groups corresponded with some interpersonal turbulence in the groups. Existing leadership structures and roles were challenged when previously quiet mothers became verbal.

There were indications that the group format may have proved difficult, initially, for the high-stress mothers, some of whom may have found the early months of group involvement to be personally uncomfortable or logistically difficult. They may have felt their group involvement was a necessary "dues payment" for the receipt of individual staff consultation, in light of the CFNP commitment to the discussion group concept. These mothers faced pressing, stressful circumstances for which the diffuse resources of a discussion group would have been difficult to tap for assistance. Staff assistance was more direct, precise, and immediate. This interpretation of the delayed integration phenomenon reflects a socio-ecological perspective.

Psychoanalytic interpretations of this finding would suggest that the initial months of infrequent attendance and minimal involvement in group discussions constituted a testing period wherein mothers were challenging the trust and commitment levels of program staff and participants (Lyons-Ruth, Botein, & Grunebaum, 1984). It also is plausible that the high-stress mothers were engaged in observational learning in the initial months of group involvement; perhaps they were monitoring how a group functions and the roles members enact.

The Child and Family Neighborhood Program was responsive to high-stress life circumstances to the extent of providing the expertise of professionals on a one-to-one basis. A good deal of this staff work involved large amounts of time, including coordination with other agencies. These services were seen as peripheral to the central program component of small group interaction, however. The program did not remove the expectation that participants would be involved in a group. Perhaps if home visitation activities *without* group involvement had been viewed as a legitimate form of program participation, the high-risk mothers would have pursued a different form of interaction with the program. We have no data to suggest that harm was done to the high-stress mothers by expecting group participation. Presumably they would have withdrawn if the involvement had proved to be excessively difficult.

While the basic structure of the intervention design (group discussion) remained invariant, there was a major shift in the *content* of group discussions over time, presumably in response to the interests of participants. Specifically, there was a clear pattern of decreased interest in parent–child topics over time in the CFNP groups. During the first 12 months, discussion of parenting topics during the 50-minute formal meeting decreased from an average of 29 minutes in the first 3 months to an average of 9 minutes in the last 3 months. There was a significant increase in discussion of topics related to home, family, and community (Powell & Eisenstadt, in press).

The amount of environmental stress in program participants' lives is not the only variable to consider in relation to program design. It appears there may have been a relationship between some aspect of the program and the number of children a participant had. An analysis of characteristics of early terminators of program involvement revealed that early terminators were more likely than long-term participants to have only one child. In addition, mothers who joined the program but then terminated their involvement early were found to have less extensive social networks and community involvements than long-term participants had (Powell, 1984).

Staff Organization and Roles

The staff of the CFNP was organized administratively with a minimum of hierarchy. There were three staff levels: the director, who was based at the sponsoring institution (The Merrill-Palmer

Institute), with responsibility for the evaluation and design of program operations; a program coordinator, who was based at the project site and had responsibility for daily operations; and the staff of paraprofessionals and professional specialists (part-time nurse and social worker) who worked directly with parents and children. In practice there was frequent communication across all levels, with no rigid lines of hierarchical communication. Friday afternoons were devoted to staff meetings, which involved all but temporary part-time staff members. Any staff member was free to contribute any item to the agenda, and indeed the paraprofessional staff participated in meeting deliberations as much as the professional staff. The physical arrangement of the setting for the meetings was a large circle, indicative of the democratic orientation toward information sharing and decision making. In the staff meetings and other program operations, the distinction between professional and paraprofessional designations was not pronounced; the titles of position (such as group leader) were used to designate staff rather than level of training.

The staff meetings were a major mechanism for organizational learning. Perceptions of situations and individuals were discussed openly and in detail. The meetings enabled all staff to pool information and collectively generate what might be called a program response repertoire. The bulk of the meeting topics dealt with the program's relations with the immediate social environment, with the primary focus on participants. At one meeting, for instance, there was considerable discussion of whether efforts should be made to limit use of the program's backyard play area to program participants. Children from the neighborhood used the area on weekends and evenings when the program was not in session. Several staff members viewed the program's play area as a facility for program participants only, while other staff thought the program was not to be a "private club" within the neighborhood but a resource for all community residents. At issue, of course, was the nature of the program's relationships with its neighbors. (Staff decided against a policy of limiting use of the play area to program participants only.) The meetings also explored issues and developments within the parent groups, and staff experiences with participants.

The original plan called for the paraprofessional staff to engage in the recruitment of participants as well as the facilitation of groups. Paraprofessional involvement in recruitment did not work well, however. An initial recruitment effort was the door-to-door canvassing of the neighborhood by staff (both paraprofessional and professional) to distribute informational flyers and talk with neighbors about the program. The experience was a negative one for the paraprofessionals. They reported feelings of resentment when residents responded with indifferent or hostile reactions to the program's presence. They also felt unprepared to deal with the range of questions and problems that might be posed in a home visit with a prospective or actual participant. Hence, it was decided that the public health nurse and social worker would take responsibility for recruitment and home visits.

The democratic character of the staff meetings reflected the autonomy and self-direction of staff activities in general. Within broad and flexible limits, the group leaders (paraprofessionals) had a high level of discretion in handling specific aspects of group life. Similarly, the nurse and social worker (professionals) enjoyed enormous discretionary power in determining the scope and frequency of home visits. To a great extent, they also decided who among the program participants would be involved in individual consultation work. As a consequence of this stance, much of the direct work with parents was implicitly off limits to supervision. On the other hand, the work *was* highly observable. The director and program coordinator periodically visited group meetings. In addition, the research staff observed groups on a frequent basis. Yet the program ethos of staff autonomy—which, as discussed below, was not unique to this intervention program—served to inhibit if not prevent directives or guidance to front-line workers about the specific substance of their activities. Such direction was especially needed regarding the group leader's control of group discussions. It was hoped that over time the group discussions would increasingly be dominated by parents, not staff. However, our quantitative analysis of group discussion, based on systematic observations, indicated virtually no change over a 12-month period in the dominance of staff in group discussions (Powell & Eisenstadt, in press).

Design Appropriateness

Through its recruitment efforts, the CFNP cast a wide net across the neighborhood. While no data are available on the size of the pool of potential participants, the extensive and multi-method approach to finding mothers of infants prompts the impression that probably less than 10 percent of eligible prospective participants were not known to CFNP staff. The wide recruitment net seemed to secure a distinctive subset of mothers, however. It did not secure mothers who felt superior to program participants in terms of parenting skill or socio-economic status. At the other extreme, the program did not serve large numbers of mothers from transient families that we assume were of a high-risk nature. The program *did* involve a substantial number of mothers experiencing high levels of environmental stress, but whose life circumstances were organized to the extent that they could incorporate the structure of twice-weekly group meetings. They knew the day of the week and could be ready for the program van to pick them up at the appointed time. They also possessed adequate social skills and personal confidence to join a small group of relative strangers.

The Child and Family Neighborhood Program did not realize the original expectation of serving mothers who were geographically separated from their kin, which was the image of most mothers in the neighborhood that was held by local officials at the program's onset. Seventy-five percent of the parents in the program were born in the greater Detroit area, and for more than ninety percent of the participants, at least one parent lived in the same geographic area. This is not to suggest that the neighborhood did not contain persons who were physically separated from their kin. Perhaps these individuals were part of the highly transient population to which the program could not connect. Unfortunately we have no data on the social networks of *all* mothers living in this neighborhood.

In developing the program design, staff considered the extremes of short-term classes versus a long-term group, as noted earlier. In practice, the continuous group format used by CFNP did not provide easy exit and reentry opportunities for participants. Vacation times were the only convenient periods for withdrawal. Moreover, the long-term group format proved to be problematic when participant attrition led to small group sizes. Groups with dwindling membership were not an efficient use of staff time, yet they often showed understandable resistance to the influx of new members or a "merger" with another group. Perhaps a built-in periodic break (for instance, one every four months) would have provided a structured ending and beginning mechanism for facilitating departures and entrances.

It was surprising to learn that first-time mothers were more likely than multiparous mothers to terminate their involvement early. This runs contrary to the view that first-time parents are in the greatest need of parent programs. The Child and Family Neighborhood Program included a preschool program for siblings while mother and infant were involved in group meetings. Initially this service was seen as an auxiliary component. Hindsight indicates that perhaps it was one of the chief attractions of the program. Mothers frequently indicated that they came to group meetings because of the older child (for example, "He was up early, standing at the window and watching for the van to come and pick us up."). Having a child involved in the preschool program also gave mothers another set of contacts with program staff (preschool workers) and a wider range of topics to discuss with fellow participants (such as, "How does your kid like the preschool?") than first-time mothers had. We have no anecdotal or systematic data to suggest that first-time parents felt less connected to the program than multiparous mothers. Participation in a homogeneous group of first-time parents rather than the mixed arrangement used in the CFNP may have increased the retention of first-time mothers in the program.

IMPLICATIONS FOR PROGRAM DESIGN AND RESEARCH

Program Design

To what degree can a program be fully responsive to a community? The CFNP experiences suggest several sources of limits to program responsiveness. On the surface the low-income neighborhood appeared to be a homogeneous setting. A closer look revealed at least three distinct groups of potential clients that CFNP was unable to serve simultaneously. The factors that appeared to keep CFNP from realizing the lofty goal of serving all people included the program commitment to the peer group format, seemingly incompatible values among prospective participants, and limited program resources.

As noted earlier, CFNP never altered significantly its basic commitment to the peer group format. The provision of professional staff consultation as part of client recruitment represented an unplanned program response to the realities of neighborhood residents. But individual consultation through home-based work at the recruitment stage was an explicit means to an end: Help was offered as preparation for and a staff link to the peer group. One can imagine alternative program designs in light of the aforementioned characteristics of the potential client pool. One strategy would have been to offer two delivery systems of equal status: home visitation and peer discussion group. Another option would have been to include a drop-in center concept in addition to or in lieu of the peer groups.

The fairly rigid adherence of CFNP to the peer discussion group design reflects a critical program development issue in the field: To what extent should client characteristics determine the nature of services offered? Brookfield (1986) has been particularly adamant in suggesting that professionals should not abandon their notions of what clients need. In the case of CFNP, the "professional knows best" orientation was manifested chiefly in the *method* or structure of the intervention (i.e., long-term peer group), not the content.

The incompatible values and life-styles of the potential client subgroups uncovered by CFNP proved to be difficult to accommodate. No doubt this occurs in most programs using a center-based mode wherein individuals of different backgrounds are expected to form interpersonal ties. The Child and Family Neighborhood Program made little effort to accommodate the "have" mothers who decided not to associate with the "have-not" participants in the program. Presumably a staff screening system could have been implemented to identify and then cluster mothers of similar values. Blatant forms of client tracking were opposed by most CFNP staff. As noted earlier, it was our impression that most mothers who put themselves in the "have" category perceived the program as an operation for "have-nots" even before or without meeting anyone at the program. Stereotypic images of parent education and support programs need careful consideration in determining the messages to be communicated in program recruitment activities.

Financial resource limitations would have been a deterrent to the development of ongoing home-based work if CFNP had chosen to move in this direction. The home-based work that was provided by CFNP represented a considerable expenditure of staff energies. Like most interventions, CFNP had insufficient funds for responding to all needs of parents in the community. At the same time, it is highly likely that CFNP would have significantly altered the group format to free up resources for more home-based work if a sufficient number of individuals had not been secured for the discussion groups.

Staff Organization

Existing research on the administrative organization of staff in parent–child intervention programs is too scant to suggest empirically derived generalizations on how best to organize communication and decision-making processes within a program. Similar to the CFNP experience reported herein, process studies of parent–child intervention programs point to staff autonomy and control over the content of services rendered (Mindick, 1986; Travers, Irwin, & Nauta, 1981).

A dilemma seems to be operating here. On the one hand, new programs must move quickly to learn about their environments and respond in adaptive ways. A program's ability to engage in this early and necessary organizational learning presumably is improved when communication easily flows upward as well as downward. As noted at the outset of this chapter, the organization behavior literature points to the value of open communication channels when there is environmental uncertainty. Added to this is the commitment of many intervention programs to the principle of staff autonomy and control due to one or more of the following theoretical rationales: (a) discretionary power is needed by staff to respond to individual circumstances and needs (Hasenfeld & English, 1974); (b) self-directing staff provide a role model for program participants; and (c) staff autonomy and participation in program decision making contribute to higher worker satisfaction and productivity (Cherniss, 1980). On the other hand, open communication channels may lead to decision making by group consensus or some other form of committee or task force structure that includes a representative group of workers. Vital information about the environment is unlikely to flow freely up a program's hierarchical ladder if the information is acted on in ways that ignore the involvement and wishes of the information provider. The problem with group decision making is that it often is inefficient, requiring time a new program usually cannot afford to provide. Moreover, as happened with CFNP and other interventions, staff discretionary power may lead to approaches or behaviors that conflict with program intentions or desired ways of working with parents. Supervision does not come easily when the operational norm calls for staff to be trusted to provide whatever type of support a parent needs most.

Resolution of this dilemma depends on further experimentation with alternative models of organization, and research on the consequences of staff communication and decision-making patterns for worker behavior and attitudes. The responses of paraprofessionals and lay volunteers to roles involving higher amounts of discretionary judgment are especially in need of investigation. The CFNP paraprofessionals seemed most uncomfortable in situations with high ambiguity and unpredictability surrounding client needs and appropriate staff responses. To reduce the ambiguous nature of these situations by imposing a structured curriculum or treatment plan would run contrary to the basic reason for allowing staff judgment to determine what clients need. The question here is what types of personal characteristics and organizational structures are predictive of effective paraprofessional functioning in an intervention program.

Research Considerations

This chapter discussed a type of inquiry focussing on the program as the unit of analysis and hence is a departure from the typical research interest in participant outcomes. There is an important conceptual issue in determining how to measure program interactions with potential and actual clients. Do we use individual subjective perceptions of clients, or more objective measures? With the program as the unit of analysis, it can be argued that the most valid indicator of program–client relations is how program staff perceive the situation and clients. This position reflects the phenomenological research tradition and is represented in Bateson's (1972) idea that behavior is dependent on the meaning of events rather than the events themselves. Client impact on program design is thought to depend on staff perceptions of clients. With this view, alterations in a program's relationship to clients are a function of shifts in staff members' perceptions of the clients, and not changes in clients *per se.*

This issue connects to the research practice of using self-report data on program functioning from two or more individuals. A critical issue is whether differential reports are to be viewed as valid perceptual discrepancies or as measurement error. Efforts to compare and contrast various constructions of a program and its environment seem consistent with a conception of the program as a dynamic, changing entity. To approach discrepant reports as error variance implies that there is a one "true" representation of reality. This conceptual issue seems to get lost in some methodological discussions of the number of system members needed for an adequate investigation of a system. For instance, one can argue that in studies using self-report measures, researchers should include more than one person in the system as a way of generating corroborative evidence. While the experiences and reports of one person cannot be generalized to others or be used to characterize the program, our understanding of programs and their clients is reduced to a simplified level when different responses from two or more persons are treated as a "check" on measurement instead of a valid perceptual difference between members of the same program.

CONCLUSION

In the field of community-based interventions, an ecological perspective on program development requires serious examination of program–environment interactions. Potential and actual clients are a major component of a program's social context. The assumptions programs make about what individuals need are a fruitful area for research and analysis. The present chapter suggests that program structure and content provide a particular weave in an intervention net that, when cast across a community, is likely to attract a distinctive set of program users. How programs accommodate the responses of potential and actual users to a particular design is an equally important area of inquiry. This chapter points to several sources of limitations on program responsiveness to client characteristics. A key issue to be addressed is the process by which program designers attempt to reconcile differences between theories or intuitions regarding appropriate intervention designs and the characteristics of desired users of the program. Determining the best match between program and participant is increasingly of major interest in the human services field. Careful attention to the interplay between client characteristics and program design is likely to enhance the process by which we match programs and people.

REFERENCES

Argyris, C. & Schon, D. A. (1978). *Organizational learning: A theory of action perspective.* Menlo Park, CA: Addison-Wesley.

Bateson, G. (1972). *Steps to an ecology of mind.* New York: Ballantine.

Brookfield, S. (1986). *Understanding and facilitating adult learning.* San Francisco: Jossey-Bass.

Burns, T. & Stalker, G. M. (1961). *The management of innovation.* London: Tavistock.

Cherniss, C. (1980). *Staff burnout: Job stress in the human services.* Beverly Hills, CA: Sage Publications.

Chilman, C. S. (1973). Programs for disadvantaged parents. In B. M. Caldwell & H. N. Ricciuti (Eds.), *Review of child development research,* Volume 3 (pp. 403–65). Chicago: University of Chicago Press.

Eisenstadt, J. W. & Powell, D. R. (1987). Processes of participation in a mother–infant program as modified by stress and impulse control. *Journal of Applied Developmental Psychology, 8,* 17–37.

Hasenfeld, Y. & English, R. A. (1974). *Human service organizations.* Ann Arbor, MI: University of Michigan Press.

Katz, D. & Kahn, R. L. (1966). *The social psychology of organizations.* New York: John Wiley.

Laosa, L. M. (1984). Social policies toward children of diverse ethnic, racial, and language groups in the United States. In H. W. Stevenson & A. E. Siegel (Eds.), *Child development research and social policy* (pp. 1–109). Chicago: University of Chicago Press.

Lewin, K. (1958). Group decision and social change. In E. Maccoby, R. Newcomb, & E. Hartley (Eds.), *Readings in social psychology,* 3rd ed. (pp. 197–211). New York: Holt, Rinehart & Winston.

Lyons-Ruth, K.; Botein, S.; & Grunebaum, H. (1984). Reaching the hard-to-reach: Serving isolated and depressed mothers with infants in the community. In B. Cohler & J. Musick (Eds.), *Intervention with psychiatrically disabled persons and their young children* (pp. 95–122). San Francisco: Jossey-Bass.

Miles, R. H. & Randolph, W. A. (1980). Influence of organizational learning styles on early development. In R. H. Miles & W. A. Randolph (Eds.), *The organizational life cycle: Issues in the creation, transformation, and decline of organizations* (pp. 44–82). San Francisco: Jossey-Bass.

Mindick, B. (1986). *Social engineering in family matters.* New York: Praeger.

Powell, D. R. (1984). Social network and demographic predictors of length of participation in a parent education program. *Journal of Community Psychology, 12,* 13–20.

Powell, D. R. (1987). A neighborhood approach to parent support groups. *Journal of Community Psychology, 15,* 51–62.

Powell, D. R. (in press). Processes of participation in support groups in a low-income neighborhood. In B. Gottlieb (Ed.), *Support groups.* Beverly Hills, CA: Sage Publications.

Powell, D. R. & Eisenstadt, J. W. (in press). Informal and formal conversations in parent education groups: An observational study. *Family Relations.*

Rogler, L. H.; Malgady, R. G.; Costantino, G.; & Blumenthal, R. (1987). What do culturally sensitive mental health services mean? The case of Hispanics. *American Psychologist, 42,* 565–70.

Schaefer, E. S. (1977). *Professional paradigms in programs for parents and children.* Paper presented at the annual meeting of the American Psychological Association, San Francisco.

Thompson, J. D. (1967). *Organizations in action.* New York: McGraw-Hill.

Travers, J.; Irwin, N.; & Nauta, M. (1981). *The culture of a social program: An ethnographic study of the Child and Family Resource Program.* Cambridge, MA: Abt Associates.

8

BETWEEN CAUSE AND EFFECT:
THE ECOLOGY OF PROGRAM IMPACTS

MONCRIEFF COCHRAN

The Family Matters Project, coordinated by Moncrieff Cochran, was one of the earliest (1976) and remains one of the largest family support interventions to be undertaken using an explicitly ecological model. Jointly developed by Cochran and Bronfenbrenner, the Project embraces a large number of ecological themes and issues. Cochran highlights the importance of better understanding the interactions that take place within and between systems and the value of ecological approaches vis-à-vis more traditional interventions.

In this chapter I shall use the Family Matters Project as an example of assumptions made and methods applied when undertaking an ecological approach to the provision of support to families, and to the evaluation of such an effort. The chapter begins with a summary of the purposes of our parental empowerment program, and an outline of what we view as its "ecological" dimensions. This is followed by a brief presentation of program content, and an overview of the research strategy used to evaluate program impact. The remainder of the chapter is devoted to a discussion of how the underlying assumptions and the methods employed in our ecological orientation affected what was learned about the effects of the program.

When the Family Matters Project was first formulated by Urie Bronfenbrenner and myself in 1976, it had several overlapping purposes. One was to develop and implement a program of family supports for parents and their young children based on the assumption of strengths rather than deficits. The program was intended to give positive recognition to the parenting role; stimulate the exchange of information with parents about children, neighborhood, and community; reinforce and encourage parent–child activities; encourage mobilization of informal social supports; and facilitate concerted action by program participants on behalf of their children. Another purpose had a more general aim: to understand better what constitutes "resources" to adults responsible for raising their own children. Finally, we were interested in the program as a way of nudging the social and psychological adaptations made by parents to their particular life circumstances, in the hope that responses to such a stimulus might cast in sharper relief the key features of family ecologies and contribute to our scientific understanding of family life (Bronfenbrenner & Cochran, 1976).

The Family Matters program was designed with explicit reference to an ecological perspective. This perspective was manifested in a number of interlocking ways. First, the program paid particular attention to systems outside an individual's psychic processes, in particular the parent–child microsystem, and two systems operating at the meso-level—the parents' personal social networks and communication linkages between home and school (Bronfenbrenner, 1979; Cochran & Brassard, 1979). Of special concern were the roles played by parents in mediating the influences of larger systems (network, neighborhood, school, workplace) on their child's development. Second, the program was delivered to a variety of kinds of families, and was made flexible enough to accommodate various expectations and needs, because of our particular interest in comparing the differing environmental circumstances faced by those groups of families typifying urban America. Our particular concern in programming for the parents' definitions of appropriate subject matter and developmental goals stems from the phenomenological orientation that underlies much of the past and present thinking associated with the ecology of human development and family life (Bateson, 1972; Bronfenbrenner, 1979; Mead, 1934).

These theoretical starting points also both reflected and influenced the assumptions underlying the family supportive process that came to be known (largely in retrospect) as the parental empowerment process. We assumed from the beginning that *all* families have strengths and that much of the most valid and useful knowledge about the rearing of children can be found in the community itself— across generations, in networks, and in ethnic and cultural traditions—rather than in the heads or books of college professors or other "experts" (Berger & Neuhaus, 1977; Ehrenreich & English, 1979). We also recognized the legitimacy of a variety of family forms, the important contributions made by fathers to the parenting process, and the special value in cultural differences. The details of the parental empowerment program have been presented elsewhere (Bo, 1979; Cochran, 1985; Cochran & Woolever, 1983; Mindick, 1980; Mindick & Boyd, 1982). Here we shall limit ourselves to a review of the basic goals underlying the program, and the processes engaged in to achieve those goals.

THE FAMILY MATTERS PROGRAM

The goals of the program were all related broadly to the parenting role, and ranged, on a parent-involvement continuum, from simple engagement and awareness to more active initiation and follow-through. In the first instance, the aim was to find ways to recognize parents as experts, based on our assumption that parents brought strengths and special expertise to child rearing and our awareness of the systematic ways in which such recognition is provided to parents in other cultures (Kamerman & Kahn, 1981). Another goal was to exchange information with family members about children, the neighborhood, community services, schools, and work. Here we were responding to the body of literature (Caplan, 1974; Sarason et al., 1977) identifying resource exchange as a key to the maintenance of mentally healthy communities. Reinforcement of and encouragement for parent–child activities was a third goal of the program, and this priority stemmed from the recommendations of those reviewing the education programs of the 1960s and early 1970s, who concluded that active involvement of parents in the learning of children was a key to success (Bronfenbrenner, 1974; Florin & Dokecki, 1983). A fourth goal involved social exchange beyond rather than within the immediate family: the exchange of informal resources such as baby-sitting, child-rearing advice, and emotional support with neighbors and other friends. This informal exchange process was distinguished from the information and referral process more commonly associated with formal agencies and community organizations (Cochran & Brassard, 1979; Collins & Pancoast, 1976; Gourash, 1978; Killilea, 1976; Stack, 1974; Tolsdorf, 1976). Finally, we wished to facilitate concerted action by program participants on behalf of their children, where those parents deemed such action appropriate. A neighborhood-based community development process was envisioned, in which needs assessments carried out by the parents of young children would lead to the identification of issues of common concern and to a change in efforts related to those issues.

Implementation Strategies

The program was offered to 160 families, each containing a 3-year-old child, in 10 different Syracuse neighborhoods. Initially, two separate mechanisms were used to involve families in activities related to their children. One, a home-visiting approach, was aimed at individual families and made available to all participating families in half of the program neighborhoods. A great deal of emphasis was paid in home-visiting to finding ways of providing positive recognition to parents for their contributions in the parenting role. This building of self-esteem often led to needs assessments initiated by the parents themselves, and thereby to a broadened facilitator role for the home visitor (Cochran, 1985).

Families in the other five neighborhoods were asked to become involved in group activities with clusters of other Family Matters families in their own neighborhoods in an effort to emphasize mutual support and cooperative action, with family dynamics and the parent–child dyad as a secondary (although still explicitly acknowledged) focus. Child care was provided at all cluster-group gatherings, and the content of the sessions included socializing as well as group activities aimed at finding solutions to neighborhood problems of common concern.

We had predicted in our original grant proposal (Bronfenbrenner & Cochran, 1976) that a combination of home visits and clusters would be more attractive to parents than either approach alone. Two early findings seemed to confirm that hypothesis. On the one hand, once certain families became comfortable with home visiting they began to express an interest in meeting neighbors involved with the program, forcing workers into the difficult position of having to resist the constructive initiatives of parents in order to prevent contamination with the cluster-building approach. On the other hand, only about half of the invited families in the cluster-building neighborhoods could be coaxed out of their homes and into group activities.

Based on these two sources of programmatic tension, we decided after nine months to merge the two approaches. Workers in the group-oriented neighborhoods began to make themselves available as often as every two weeks for home visits that focussed initially on parent–child activities, and workers who had been doing only home visits started to facilitate the formation of neighborhood groups and clusters. One consequence of access to both components of the newly integrated program was an increase in overall program participation. Initially this took the form of more parent–child activity home visits, primarily to families who previously had been offered only the neighborhood linking alternative. With more time came involvement by more families in clusters and groups; some participated simultaneously in both home visiting and neighborhood-based group activities.

As the children associated with the program grew older and approached the age of entry into kindergarten and first grade, we placed increased emphasis on programming related to the transition from home to school. These activities, prepared for delivery in both home-visiting and cluster-grouping formats, focussed on topics like the values of home and school, evaluating kindergarten and first-grade classrooms, preparing for a parent–teacher conference, understanding the child's report card, and parent–child activities for school readiness. The emphasis in each of the activities was always on the parent as the most important adult in the life of the developing child.

Families were involved with program activities for an average of 24 months, and the program itself came to a close early in the summer prior to first-grade entry for most of the target children included in the study.

EVALUATION OF PROGRAM IMPACT

Sample Design

In the design and selection of a sample for this study, we set out to accomplish several competing objectives. First, there needed to be enough families to permit inclusion of a broad range of family types, thus permitting some generalization of findings and the study of reasonably detailed distinctions among families and individuals, where indicated by the data. Second, and acting strongly to limit the first objective, we wished to utilize a relatively time-consuming, in-depth interviewing procedure, in order to obtain the kind of detailed case material that makes possible the qualitative search for statements of causality as well as broad-scale quantitative examination of relationships.

Neighborhood Selection.   We employed a stratified random sampling procedure at the level of both neighborhoods and families. First, 29 city and 28 suburban neighborhoods in the Syracuse, New York area were identified. The neighborhoods were then further classified by income level and by ethnic/racial composition. Using 3 income levels and 4 ethnic/racial levels, we randomly selected neighborhoods within the 12 subclasses (where such neighborhoods existed), giving a total of 18 main-study neighborhoods.

Selection of Families. Once study neighborhoods had been specified, we began the process of identifying all the families with a three-year-old child in each neighborhood. Race (black vs. non-black), family structure (married vs. single), and sex of target child were factors of primary interest. We then employed a stratified random sampling method within each neighborhood, choosing families within each of the eight subgroups defined by family race, family structure, and sex of child, and oversampling for Afro-American and single-parent families. This method of sampling resulted, as was our intention, in a higher proportion of black and single-parent families than in the Syracuse area as a whole, and also ensured a substantial sample of ethnic white families.

Stratifying by the variables discussed above, including neighborhood income, resulted in a good sample distribution across family income. Employment status of the mother was also well distributed; approximately half of the women in our study were working outside the home (some part and some full time).

The rate of agreement to participate varied by neighborhood, ranging from nearly 100 percent in certain neighborhoods to approximately 50 percent in others.[1] Table 8.1 shows the distributions of families that participated in the follow-up phase of the research and constituted the sample available for pre–post comparison of program effects.

Sample Attrition. The possibility of differential patterns of attrition in the treatment and control groups from Time 1 to Time 2 was considered in some detail (Cochran & Henderson, 1985). Analyses indicated no attrition differences by program assignment or other factors.

Operationalization of Ecological and Psychological Domains

Our conceptual schema is presented in Figure 8.1. It provides an overview of the hypothesized relations among the major classes of variables. Home–school communication and the child's perform-

---

[1]Refusal rates by race, family structure, and sex of child are shown in Cochran and Henderson, 1985.

TABLE 8.1   Number of Families by Program, Race, and Marital Status

		Control	Program	Total
Black	Single	19	21	40
	Married	10	13	23
White	Single	16	23	39
	Married	<u>54</u>	<u>69</u>	<u>123</u>
	Total	99	126	225

ance in school, although conceptually distinct, are shown in a single box, to minimize the number of connecting arrows. The same is true for socio-demographic variables and the program. The variables presented in this chapter consist of a subset of those described in the final report to the National Institute of Education (Cochran & Henderson, 1985), selected because they provided the most insight into the program–control comparisons considered there. In the case of *mothers' perceptions of themselves as parents* the variable consisted of the mother's rating of her performance on a 25-item checklist, with each item consisting of a 7-point scale. The four *mother–child activity* variables—talk, creativity, tasks, and companionship—were derived from a set of 55-checklist questions completed by the mother, each of which was presented as a 4-point scale. The *social network* variables were concentrated in the primary network. They included change in number of primary ties between baseline and follow-up (both kin and nonkin) and number of kin and nonkin found in the primary network at follow-up who were nowhere present in the network at baseline ("new primary membership"). The *home–school contact* variables consisted of estimates by both parents and teachers for the numbers of conferences, notes, and telephone calls initiated from the home and from the school. Finally, the *school outcome* variables were drawn from the teacher questionnaire and included the following domains: personal adjustment, interpersonal relations, relationship to teacher, cognitive motivation, and report card score averaged across core subjects.

Summary of Statistical Methods

The core of our first stage statistical analyses involved single-equation models, using regression techniques (including analysis of

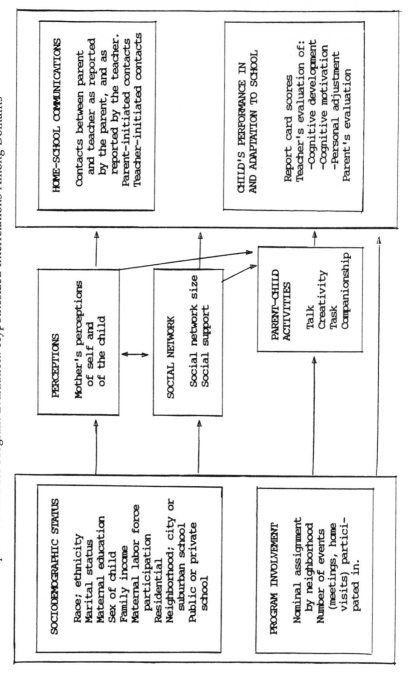

FIGURE 8.1 Conceptual Schema for Program Evaluation: Hypothesized Interrelations Among Domains

variance and covariance). These models frequently involved the specification of different regressions for each subgroup in the model (analysis of homogeneity of regressions), random as well as fixed factors (mixed models), the simultaneous examination of group (ecological) and individual effects, and repeated measures on the dependent variables.[2]

Stage two analyses reflected a shift from initial interest in direct effects to concern with relations between "intervening" and outcome variables. Instead of concentrating on comparisons of means in analyses of covariance, with program assignment as the independent variable and one or another ecological outcome on the dependent side, the interest was in the homogeneity by program assignment of the regressions of one ecological domain on another. For instance, is the relationship between a change in networks over time and school performance different for families involved with the program than it is for those in the control group? The results generated for appropriate subsamples were control-program comparisons of regression coefficients representing relationships between pairs of ecological domains.

SUMMARY OF FINDINGS

This summary can only begin to introduce the reader to the range and diversity of findings that have emerged from evaluation of the Family Matters program. It is devoted primarily to an exploration of *processes* through which the empowerment program might have affected outcomes of interest, because these processes have been neglected in most evaluations of social programs and because knowledge of process builds on our ecological model and contributes to an understanding of empowerment.

Composites created from the findings presented elsewhere (Cochran & Henderson, 1985, 1986) are shown in Figures 8.2 and 8.3. Program impacts *directly* related to each of the ecological fields of interest are shown as lines connecting the program with each of those fields. Of even greater interest to us is how involvement with the program might have affected *relations between* ecological fields— the link between social networks and perceptions of self as parent. Relationships between pairs of these domains were examined as a function of exposure to the empowerment program, controlling as

(*text continues on p. 154*)

---

[2]For more detail, see Cochran and Henderson, 1985.

Black, Single Parents

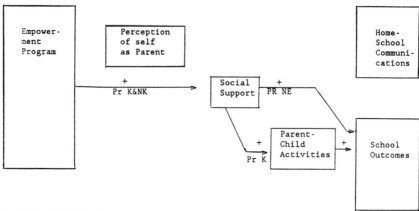

Black, Married Families

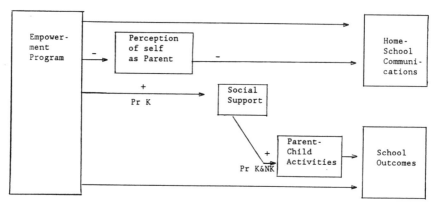

<u>Key to abbreviations</u>:

        Pr K - Primary network Kin
        Pr NK - Primary network Nonkin
        Pr K&NK - Primary network Kin and Nonkin
        NK - Network Nonkin
        K&NK - Kin and Nonkin
        Pr K - Primary network Kin
        New Pr NK - New Primary network Nonkin
        Ed - Effect was observed only for mothers with education
           beyond high school

FIGURE 8.3 Program Impacts: White Families and Links Between Model Components

White, Single Parents

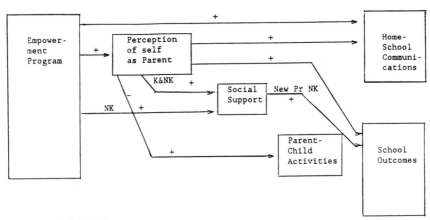

White, Married Families

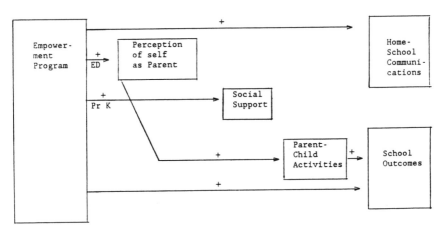

<u>Key to abbreviations</u>:

Pr K - Primary network Kin
Pr NK - Primary network Nonkin
Pr K&NK - Primary network Kin and Nonkin
NK - Network Nonkin
K&NK - Kin and Nonkin
Pr K - Primary network Kin
New Pr NK - New Primary network Nonkin

before for preexisting differences in socio-economic status. Where differences by program assignment were found in these relations between domains, they are shown in the figures as lines connecting the domains. In looking at these more complex differences between program and control samples, we believe that light is being shed on *indirect* effects, that is, processes mediating the relationship between program involvement and ecological fields relatively distant from the parent (such as the school).

The pictures provided in the figures for the single-parent subgroups suggest as a hypothesis that the impacts of the empowerment program on children's school performance are heavily mediated by changes occurring within and around their parents. In the case of the *black one-parent families,* increases in the number of relatives included in the mother's primary network were associated with reports of more joint activity with the child. Joint activity involving household chores was linked in turn with higher performance in school. And expansion of nonkin membership in the primary networks of those mothers was linked with their children's school outcomes, especially when those outcomes involved school readiness (personal adjustment, interpersonal relations, relations with the teacher). *White single mothers'* perceptions of themselves as parents appeared to be a key determinant in whether positive performance was seen in the more distant reaches of their ecological fields. Higher parental perceptions are associated, for these mothers, with expansion of their primary networks, the activities they reported engaging in with the child, their level of communications with the child's teacher, and the teacher's report of the child's progress in first grade. There is evidence that the nonkin sector of the primary network may also play a positive role in its own right, with increase in nonkin involvement with the mother linked to higher school outcomes, again primarily in the area of school readiness.

The pictures in the two figures are more ambiguous for married mothers and their children. A somewhat lower self-perception as parent by *black married mothers* in the program seemed to be tied to greater communication with the teacher in those instances where the child was perceived as having difficulty in school. There was also a direct positive link between program involvement and increased home–school communication. For these same mothers, increased involvement with kinfolk was related to greater amounts of mother–child activity. However, none of these hypothetical chains could be linked to better performance of the child in school. School performance was tied directly to program involvement, without any intermediate links to other ecological fields.

One set of possible mediating links did emerge for *white married mothers,* if those mothers had schooling beyond high school. The proposed sequence involves increased perception of self as parent, more mother–child activities, and better performance by the child in school. The reader can see from Figure 8.2 that there was also a direct link between program involvement and school performance for the children in this subgroup. These alternative routes can be tested in models specifying simultaneous equations, which will be a next step in our analysis of these data. Another step in probing for mediating factors in the worlds of these two-parent families will be to examine the involvement of the fathers in the workings of those families and interactions with the children, to see whether some aspect of that involvement helps to determine how the children perform in school.

One other aspect of the results reported to the National Institute of Education deserves mention as a prelude to later discussion. Time and again, as we compared the slopes of the regressions of one ecological subsystem on another for the program and control groups, we found a moderately positive regression line for program families, contrasted with a rather more sharply negative slope for control families. In terms of program impact, these contrasts strongly suggest that this empowerment program has *prevented* more than it has *enhanced*; the somewhat positive relationships seen for program families become much more significant when compared with the negative relationships visible in the data collected from the control families. The implications of these findings for how family support programs are conceptualized are discussed below.

LESSONS LEARNED
BY USE OF AN ECOLOGICAL APPROACH

Much more was learned from our effort to apply an ecological approach to the provision and evaluation of a family support program than can be included in this short chapter. Here I will illustrate with some central themes the ways that both the *assumptions underlying* our version of the ecological framework and *the framework itself* shaped both the direction of our inquiry and the ways that we have interpreted our findings.

Family Support as Relief from Stress

The traditional expectation associated with an intervention designed to affect outcomes in children has been that the children receiving the special treatment will then perform better than an equivalent control group. Historically the assumption underlying such a model has been that the intervention was *compensating for* some *deficiency* in the child's life circumstances that would otherwise limit performance (Dokecki & Moroney, 1983). From our ecological orientation we developed an alternative to this standard stance, based on our assumption that *all* families have strengths and experience stress. This assumption has, in turn, led to an interpretation of our findings in which the intervention is thought of as *preventing the loss* of certain family or environmental functions and therefore making possible the maintenance of child performance at an acceptable level. The assumption is not that there is a deficiency that needs correction, but rather that a system capable of functioning adequately deserves protecting. The concept of supporting the family, or family supports, is based on this second model. From this perspective the family is viewed as a system that, if given an opportunity to function in a relatively supportive environment, can fulfill the basic developmental needs of its members. Public policies designed to provide family support aim, through stress reduction, to allow families to function effectively rather than to "correct" their "deficits."

If the purpose of family support is to prevent loss of family functioning, then one would expect there to be instances in our data in which no change in the program group was accompanied by *decreases* for control families. The first example of this sort appeared in the relationship between reductions in family income and the child's performance in school.[3] For control families, lower incomes were associated with poorer school performance, in all groups except that containing married white families. This income-related decrement did not appear for the program families in these three subgroups, suggesting to us that participation in the program buffered those families against the effects of reduced income. This impression was reinforced by indications in the social network data that financial support from network members had eroded somewhat less for white single mothers in the program than for those in the control group. It was underscored yet again in the relationship of

---

[3]Detailed findings are discussed in Cochran and Henderson, 1985, 1986.

networks to school outcomes, parental perceptions, and home–school communications, where the pattern for white single mothers involved strong negative regressions for the control group, balanced by flat or moderately positive ones for the program group, after adjusting for mothers' educational level. This suggests, in the case of networks, that the program did more than simply increase the numbers of nonkin in the networks of these mothers: It also seemed to affect how those special nonkin were *brought to bear* on other aspects of family life. The impression that accumulates from these data is that the strong positive *direct* associations between program involvement by white single mothers and both their perceptions of themselves as parents and the support they received from close friends served to buffer the child against problems in school. The dynamics of this buffering process are only conjecture at this point, but a clue may be provided by the indication that when their children showed signs of having difficulty in school, those same mothers were also found to be in regular contact with the children's teacher. The general point is that interventions preventing a significant loss in the functioning of family members should be viewed with as much interest as those that produce gains in performance relative to control-group families. In fact, one can argue that the preventive role is the more important one if it is accomplished by strengthening the family rather than usurping its role and functions. It is important to reiterate that a belief in the strengths of families underlay our ecological orientation. Whether this assumption can be applied to the ecological perspective in general remains to be seen.

The Importance of Distinguishing Among Ecological Niches

Another assumption basic to our ecological orientation was that a variety of family forms are legitimate, and cultural differences are to be valued. We set out to design some of these contrasts into rather than out of the research (Bronfenbrenner, 1979), and remained alert to the possibility of other such contrasts where they occurred naturally (mother's educational level, mother's occupational status).

Of the two most basic stratifications in our sample, race and marital status, the latter was clearly the more powerful in explaining differences. This power was especially evident in our search for processes related to the transmission of program effects. These data strongly suggest that couples are able to bring their program experiences directly to bear on the school-related support of their children, while for unmarried (usually single) parents such support seems to be contingent on the interim building of self-confidence and/or social network supports.

One feature common to all of the subgroups for which positive school effects were found was their less advantageous position in the social structure. Single mothers almost invariably have fewer educational and monetary resources than do married mothers, and our sample fit this more general pattern. Positive school outcomes were associated with the children of less educated parents, whether from two-parent or one-parent families. This finding held for both Caucasian and Afro-American children.

The question of how best to treat a variable like mother's educational level (or family income) can be linked directly to the ecological perspective. Because of its continuous nature, it is tempting to use such a variable simply as a covariate, and thereby "control away" its effect in order to permit a better understanding of the impacts of categorical variables of primary interest (such as race and family structure). By taking this approach, however, one defines mother's educational level as ecologically "uninteresting," when in fact the systems approach central to the ecological orientation provides an opportunity to develop an understanding of what more or less education *means*, in access to network ties and relations with community institutions as well as in parent–child and spousal relations.

The reader will be aware by now of our preoccupation with ecological contrasts, and our conviction that much can be learned about attitudes and behavior through comparison based on these contrasts. Of course there are also costs associated with analytic designs that rely heavily on subgroup comparisons, especially if those subgroups are defined by the intersections among a number of structural dimensions (race, number of parents, sex of parent or child, occupational type, program–control, etc.). With an increase in the number of subgroups attended to comes a decrease in the size of each subgroup, until an overall sample of 200–300 is reduced to comparison among subgroups containing 10–15 families. This problem is easy to anticipate and could be overcome simply by increasing the size of the overall sample, were it not for the human ecologist's simultaneous commitment to the subjective perception of the research participant and to the understanding of cause and effect as perceived by the participant. (See also the chapter by James Anglin, this volume.) We responded to this commitment by engaging each of the parents in our study in a series of in-depth interviews, which are expensive both to conduct and to analyze. Reliance on such methods mitigates against expansion in sample size. More sensible is an approach that combines what might best be thought of as a set of small-scale, in-depth exchanges with the members of families living in distinctly different circumstances, with a large-scale survey designed to test the generalizability of both the life circumstances described in the smaller subgroups and the beliefs and attitudes expressed by the members of those subgroups. (Donna Lero argues persuasively in support of ecologically oriented surveys, elsewhere in this volume.)

The Utility of Process Variables

One contribution that an ecological orientation can make to research is to enrich our understanding of process. At the same time, the inclusion of "process" variables in a conceptual model for evaluating the impact of an intervention complicates matters at virtually every stage in the life of the project. In the case of Family Matters, reams of additional data about self-perceptions, networks, and parent–child activities had to be collected both prior to and following implementation of the program. The costs of gathering, preparing, and analyzing these data were substantial. Did the results justify the investment?

One way to answer the usefulness question is to look at Figures 8.2 and 8.3. Imagine the diagrams as they would look if they contained only the direct relationships between the program and school, and home–school outcomes. The single parents in the sample would be affected most by removal of all the links to "process" components. The impression created would have been that one-parent families had not responded to our parental empowerment approach. Beyond simply missing the fact that certain of the children in single-parent families had shown improvements in school behavior that could be associated with program involvement, removal of the "process" variables from the model virtually eliminates any opportunity to learn *what it was* about the program that seemed to make a difference to those involved with it. For instance, we invested a great deal of effort in discovering ways to give positive recognition to parents for the vitally important roles they were playing in the lives of their children. The supposition was that parents needed to feel confident about themselves as parents before they could be expected to become actively involved in the more "executive" aspects of the parenting role. The summary of findings represented by Figures 8.2 and 8.3 certainly suggests that for two of the four subgroups represented in the sample, perception of self as parent plays an active role in determining whether parents become involved with their child's teacher when there is indication that the child is having difficulty in school. While the nature of the data permits only the generation of hypotheses, the findings are nevertheless intriguing. They are also not of the simple "more is better" variety, as indicated by the fact that for married Afro-American mothers more school involvement was accompanied by a *drop* in regard for self as parent. The point is that the findings can be translated into policy at the program level. They clearly imply that white single parents will become actively involved with the teachers of their children only if they feel reasonably good about themselves as parents, which suggests that programs be designed to stimulate positive changes in such self-regard. The same kind of argument can be made for social networks and school outcomes, again especially for mothers and children in single-parent families. Such reference to specific aspects of the content of the program would not have been possible in the absence of data about "process."

Social Networks as Measures of Program Impact

This concern with what might be called "ecological processes" is no longer new. Similar conceptualizations can be found in the areas of stress and coping (Pearlin et al., 1981; Pearlin & Schooler, 1978) and family–environment relations (Powell, 1979). When the Family Matters Project was first funded in 1976, however, the idea of using a concept like "personal social networks" to embody the more generalized notion of informal support systems had not been introduced to the social science community. Even more unusual was our operationalization of the concept as a dimension of family and community life amenable to change as a function of involvement with a community-based program of family support. Inclusion of the personal network in our overall conceptualizaton as a major ecological system operating between the immediate family and societal institutions grew out of practical use of networks at the neighborhood level that I had made as a local organizer of child care systems. The formal proposal to include networks in the ecology of human development was made in 1979 (Cochran & Brassard). Thus the Family Matters research project became a proving ground for testing the validity of the networks paradigm in the ecology of family life. The evidence from our project indicates that networks have value as a way of characterizing the web of relations that links parents and children together across families and connects family members to other community settings (workplace, school, etc.). Several examples of how this dimension of the total family ecology has affected our thinking are included below.[4]

Networks and the Unmarried Mother. One indication of the importance of network ties to mothers and children can be seen in the single-parent subgroups of Figures 8.2 and 8.3. Those patterns indicate that the addition of key nonrelatives to the mother's network was associated with improved performance in school for the children of unmarried mothers, most of whom were single parents. This finding held across races. It provides insight into the needs of a family type already of numerically significant size in the United States, which, in the 1970s and 1980s, has grown considerably as a proportion of all the families with young children. While work remains to be carried out in an effort to describe more fully the key additions to these mothers' networks, the indications are that some women did not passively accept social relationships offered them through the good offices of the program, but instead were encouraged by involvement with the program to take the initiative in marshaling social resources for the many demanding tasks at hand, one of which was raising a young child. Success in recruiting such assistance seems to have had payoff both for parent (self-regard) and child (school performance). One implication of these findings is that the concept of social support for the child-rearing process should be expanded beyond the traditional spousal relationship to include the network of friends and relatives.

Kinship and the Afro-American Family. Readers should not infer that social supports are equated primarily with social ties beyond kinship. Figures 8.2 and 8.3 indicate that three of the four subgroups (defined by marital status and race) showed direct increases in primary kin ties associated with program involvement. An important characteristic accompanying this relationship was the race of the mother. Afro-American mothers were significantly more likely than Caucasian mothers to increase their involvement with primary kin if included in the program, and this carried over to unmarried women. It would be easy to dismiss this finding as an inevitable re-

---

[4]A cross-cultural comparison of mothers' networks can be found in Cochran, Gunnarsson, Grabe, and Lewis, 1984.

sult of minority status, racism, and poverty, saying that such women are forced to rely on close relatives because of limited access to social relationships with members of the white majority and the cost of maintaining social ties with nonkin. Such a view, while seeming to fit the data, is deficit driven and incomplete. We prefer the view that Afro-American families provide one of many models for carrying out the rearing of the young in our culture, and that kinship in general plays a larger role in those families than is the case for American Caucasians. This view implies that any model should be evaluated on its particular merits, and in this case some of those merits can be identified in our data. There was indication (Cochran & Henderson, 1985) that black unmarried mothers in the program received financial assistance from greater numbers of relatives over time, despite the sharp recession, while the reverse was true for the white unmarried subgroup. The findings shown in Figure 8.2 indicate that, in both black subgroups, increases in the number of primary kin reported over time were associated with larger amounts of parent–child activity. No signs of negative impact associated with kin ties surfaced to counterbalance these positive indications, leaving us to conclude that these families lost nothing, and may well have benefitted, from growth in their relationships with relatives.

Network Changes: A Good Thing?   One overall impression conveyed by our findings is that participation in the Family Matters program had the effect of moving mothers and their families toward patterns of informal social relations that they might otherwise have realized more slowly, and perhaps less fully. Were those shifts in network relations a good thing? One way to address that question is by examining the links between program-related network increases and other process and outcome variables, as portrayed in Figures 8.2 and 8.3. The network appears to be a key transmission center for white unmarried mothers, primarily through the nonkin sector, the growth of which is positively associated with perception of self as parent and the child's performance in school. Black unmarried mothers involved in the program also showed substantial growth in the network, with kin linked to increases in parent–child activities and nonkin to improved performance by the child in school. Less can be said about the impact of expanded primary kin networks for program mothers in the married subgroups, where the only link was with parent–child activities for the Afro-American portion of the sample. On balance, there is little in our data to indicate that the expansion of the primary network associated with

participation in the Family Matters program had deleterious consequences, and considerable indication of positive contribution, especially for unmarried mothers. A different set of outcome measures
might have led to an alternative conclusion, of course, but our data
leave us cautiously optimistic about the consequences for mothers
and children of facilitating network-building activities.

Because there was concern about disrupting or changing the
social ties of families participating in the Family Matters program,
the program was never advertised as designed especially for network-building purposes, nor did any impetus develop to become
especially active in that regard. Neighborhood cluster building was
an avowed goal, but it was espoused much more in the interest of
collective action on behalf of child, family, and neighborhood than
to provide parents with material and emotional support. The kinship potential in the networks was virtually ignored; we made no
effort, for instance, to encourage parents to invite relatives to home
visits or cluster-group meetings, although kinfolk did attend some
of those occasions in the normal course of events. It is fair to say that
our networking initiatives were quite passive. *Many of these findings
might be expected, therefore, to be associated with any facilitating program of family support.*

Networks as Convoy. One of the exciting aspects of social supports as program outcomes is their potential for the development of
the individual in the future as well as the present. House (1980) uses
the convoy analogy, which I also find useful. Such an analogy clearly
implies that network changes associated with the program might be
as strongly linked to subsequent developments in the child as they
are to more immediate ones. The findings reported here begin to
provide outlines for the forms of transport making up such con-
voys. One vehicle is likely to be composed of close friends and relatives committed to the welfare of both parent and child. Another is
parental self-confidence. A third vehicle, and perhaps the one to
head the convoy, is the parent's level of formal education. Contained in these conveyances are resources essential to sustaining the
child throughout the developmental journey: human energy, time,
material goods, information, skills, emotional support. Our evaluation of the Family Matters program provides evidence to bolster the
contention that some environments are more likely than others to
produce and maintain such supports in transaction with parents, and
that steps can be taken at the community level to change environments in ways that facilitate family functioning.

The Empowerment Process: Fact or Figment?

Recently we have postulated (Cochran, 1985) the existence of an empowerment *process* consisting of a series of stages, involving transactions by the individual with progressively more distant environmental systems. We propose that positive changes in self-perception (Stage I) permit the alteration of relations with members of the household or immediate family (Stage II), which is followed by the establishment and maintenance of new relations with more distant relatives and friends (Stage III). Stage IV is seen as information gathering related to broader community involvement, followed in Stage V by change-oriented community action. MacDonough (1984) has shown with Family Matters data that parents can be located at different points along such an empowerment continuum and that for the first four stages a high score on a later stage is related with high scores on previous ones.

Through this evaluation we have mapped out a rather complex set of direct and indirect relations in an effort to assess the impacts of an intervention designed to empower parents on behalf of their children and themselves. Unfortunately, a good test of the "empowerment process" hypothesis requires data that our study is not able to provide. For instance, we have no fully comparable measures of perception of self as parent at baseline and follow-up with which to determine where mothers were at baseline in relation to Stage I and whether program involvement had changed this status in ways not reflected in the control group. A second shortcoming involves the absence of any measure for the information gathering (Stage IV). A third weakness involves our current measures of Stages II (relations with household members) and V (community action). Relationships with household members involve more than parent–child activities, and community action involves more than activities related to the child's school. In both instances our data base can provide information with which to expand understanding of those processes (with wife–husband relations for Stage II and other community institutions for Stage V), but such elaborations have thus far been beyond the scope of this evaluation.

What can be said, however, is that what has been learned to date about the effects of the Family Matters program does not *contradict* the general concept of empowerment as a process involving changes in self-perception and relationship with others both immediate to and more distant from the changing person. The findings do point to the possibility that constructive change in perception of self may not necessarily be in the direction of more positive feelings, depending on the perceptual point of departure at the beginning point of the intervention. Thus, within certain limits, the change in perception itself, regardless of valence, may stimulate other action. And, for certain of the families in our sample, this change shows solid evidence of being associated with variables like parent–child activities, primary network changes, and contacts with the school postulated to occur later in the empowerment process. Future efforts using simultaneous equations may throw more light on possible pathways through the data, but much will be left to speculation nevertheless. Most important, from the standpoint of our interest in the value of the ecological orientation for the social sciences and social policy, is the fact that this approach allows, and even encourages, the researcher to think in terms of processes and so begin to cast in sharper relief those aspects of the human ecology that are central to the development of healthy individuals, families, and communities.

Entry Points for Public Policy

A number of aspects of the ecological approach to social scientific research combine to propel its practitioners into the policy arena. The partnership formed with the research participants, the sensitivity to their points of view, and the richness of the family portraits painted by the qualitatively oriented techniques used all contribute to the desire to improve the lot of those among the participant families who are perceived to be burdened unfairly by the societies and communities containing them.

Fortunately, this approach to research, which has so much potential for uncovering the ways through which social-ecological forces affect family life, also can direct policy initiatives in some useful ways. First, by identifying the mechanisms through which forces like occupational status, educational attainment, and family structure limit or expand the horizons of parents, studies geared to the ecological model provide signposts for the development of effective community programs. For instance, if adults need to feel positively about themselves before they will take action on behalf of their children, and if such initiative is viewed as crucial to effective utilization of existing community resources, then programs must be designed with the positive recognition of parents as a central element.

The second value of the ecological orientation for policy formation is the capacity it provides for avoidance of unanticipated consequences. By providing insight into the various levels of the social ecology through which an environmental stimulus is likely to reverberate, studies using the approach illuminate the variety of effects likely to accrue from an intervention, and thus permit a much fuller appreciation of probable impacts than has been possible with previous models and methods.

Perhaps the most important contribution that the ecological approach can make to policy formulation is a product of an aspect of the ecological model to which I have not given enough attention in this presentation—the part played by reciprocal and feedback effects. If, for instance, a parent brings more attention and enthusiasm to interactions with a child, then the child is likely to respond with positive attitudes and behavior toward the parent, which may in turn both stimulate still more positive parental behavior and draw positive comments from relatives and friends. Or when the teacher communicates to a mother the importance of the parent as an educator of her child, the parent is more likely to impress on the child the importance of listening to what the teacher has to say in the classroom, and may also have nice things to say about the teacher to her friends. An important by-product of these multiply determined and multidirectional effects is the variety of entry points they provide for social policy. For example, access to parents' perceptions of the parenting role can be provided through home visits or parenting classes, through neighborhood clusters, through communications by school teachers with parents, through the flex-time and parental leave policies of employers, through the content of television advertisements, and in a myriad of other ways. One approach, or combination of approaches, to the bolstering of parental confidence and commitment may be more ideologically palatable to a given community or society than another and so may be more "policy appropriate." The awareness of an array of alternative routes to the same policy goal—the strengthening of parental self-esteem, or parent–child activities, or nonkin networks, or home–school communications—increases the probability that a policy choice will be made that can be embraced by a large proportion of those whose cooperation and involvement are essential to its successful implementation.

CONCLUSION

In this chapter I have presented a research project and a program of family support, both designed with explicit reference to the ecology of family life. In describing some of the lessons about research learned through these efforts I have indicated that they flowed as much from the assumptions about families that were brought to the research effort as they did from the research paradigm itself. Always implicit in what I have said, but given less emphasis than the research issues, is the potential that exists in this approach for the development of more effective programs in support of family life. It is heartening to see the success that those using the ecological orientation are having in identifying the range and variety of healthy ways that families with different traditions and access to resources transact with the environmental systems surrounding them. This growing body of knowledge, and the commitment to human dignity associated with it, should help to make possible the provision of community support that are sensitive to the capabilities and needs of families in ways that reflect the perceptions of the members of those families.

REFERENCES

Bateson, G. (1972). *Steps to an ecology of mind: Collected essays in anthropology, psychiatry, evolution, and epistemology*. New York: Ballantine Books.

Berger, P. & Neuhaus, R. (1977). *To empower people: The role of mediating structures in public policy*. Washington, DC: The American Enterprise Institute for Public Policy Research.

Bo, I. (1979). In support of families: The pilot testing of two experimental programs. Paper prepared for the Carnegie Corporation of New York. Ithaca, NY: Cornell University Press.

Bronfenbrenner, U. (1974). *Is early education effective? A report on longitudinal evaluations of preschool programs*. Vol. 2. Washington, DC: Department of Health, Education and Welfare, Office of Child Development.

Bronfenbrenner, U. (1979). *The ecology of human development: Experiments by nature and design*. Cambridge, MA: Harvard University Press.

Bronfenbrenner, U. & Cochran, M. (1976). The ecology of human development: A research proposal to the National Institute of Education, Cornell University.

Caplan, G. (1974). *Support systems and community mental health*. New York: Behavior Publications.

Cochran, M. (1985). The parental empowerment process: Building on family strengths. In J. Harris (Ed.), *Child psychology in action: Linking research and practice*. London: Croom Helm.

Cochran, M. & Brassard, J. (1979). Child development and personal social networks. *Child Development, 50*, 601–16.

Cochran, M.; Gunnarsson, L.; Grabe, S.; & Lewis, J. (1984). *The social sup-*

port networks of mothers with young children: A cross-national comparison. Gothenburg, Sweden: University of Gothenburg.

Cochran, M. & Henderson, C. R. Jr. (1985). Family matters: Evaluation of the parental empowerment program: A final report to the National Institute of Education, Cornell University.

Cochran, M. & Henderson, C. R. Jr. (1986). Family matters: Evaluation of the parental empowerment program: Summary of a final report to the National Institute of Education, Cornell University.

Cochran, M. & Riley, D. (1985). Mother reports of children's social relations: Antecedents, concomitants and consequences. Paper presented at the Conference on Social Connections from Crib to College, held at City College of New York.

Cochran, M. & Woolever, F. (1983). Beyond the deficit model: The empowerment of parents with information and informal supports. In I. Sigel & L. Laosa (Eds.), *Changing families.* New York: Plenum Press.

Collins, A. & Pancoast, D. (1976). *National helping networks: A strategy for prevention.* New York: National Association of Social Workers.

Dokecki, P. & Moroney, R. (1983). To strengthen all families: A human development and community value framework. In R. Haskins & O. Adams (Eds.), *Parental education and public policy.* Norwood, NJ: Ablex.

Ehrenreich, B. & English, D. (1979). *For her own good: 150 years of experts' advice to women.* New York: Anchor/Doubleday.

Florin, P. & Dokecki, P. (1983). Changing families through parent and family education: review and analysis. In I. Sigel & L. Laosa (Eds.), *Changing families.* New York: Plenum Press.

Gourash, N. (1978). Help-seeking: A review of the literature. *American Journal of Community Psychology, 6*(5), 413–24.

House, J. (1980). *Work stress and social support.* Reading, MA: Addison-Wesley.

Kamerman, S. & Kahn, A. (1981). *Child care, family benefits, and working parents.* New York: Columbia University Press.

Killilea, M. (1976). Mutual help organizations: Interpretations in the literature. In G. Caplan & M. Killilea (Eds.), *Support systems and mutual help.* New York: Grune & Stratton.

Larner, M. (1985). Local moves and social networks: The changing social worlds of mothers and children in three cultures. Unpublished doctoral dissertation, Cornell University.

MacDonough, J. (1984). Development of a scale which operationalizes the concept of empowerment as a process. Predoctoral dissertation, Cornell University.

Mead, G. H. (1934). *Mind, self, and society.* Chicago: University of Chicago Press.

Mindick, B. (1980). Salt city and family matters: Supporting families in the urban environment. Report to the Carnegie Corporation of New York. Ithaca, New York: Cornell University Press.

Mindick, B. & Boyd, E. (1982). A multi-level, bipolar view of the urban residential environment: Local community vs. mass societal forces. *Population and Environment*, special issue on the urban residential environment.

Pearlin, L.; Lieberman, M. A.; Menaghan, E. G.; & Mullan, J. T. (1981). The stress process. *Journal of Health and Social Behavior, 22*, 337–56.

Pearlin, L. & Schooler, C. (1978). The structure of coping. *Journal of Health and Social Behavior, 19*, 2–21.

Powell, D. (1979). Family environment relations and early childrearing: The role of social networks and neighborhoods. *Journal of Research and Development in Education, 13*, 1–11.

Riley, D. (1985). Father involvement in childrearing: Support from the personal social network. Unpublished doctoral dissertation, Cornell University.

Sarason, S.; Carroll, C.; Maton, K.; Cohen, S.; & Lorentz, E. (1977). *Human services and resource exchange networks*. San Francisco: Jossey-Bass.

Stack, C. (1974). *All our kin: Strategies for survival in a black community*. New York: Harper & Row.

Tolsdorf, C. (1976). Social networks, support and coping. *Family Process, 15*, 407–18.

9

TUMBLER RIDGE:
AN ECOLOGICAL PERSPECTIVE
IN EVALUATING THE PLANNING
AND DEVELOPMENT
OF A NEW COAL-MINING COMMUNITY

WES SHERA

The development of an entire community represents a rare opportunity to use the ecological model in a prescriptive rather than a descriptive manner. Tumbler Ridge, a planned resource community in northern British Columbia, utilized a variety of planning models in its development. In this chapter Wes Shera outlines those models and considers the development of Tumbler Ridge from an ecological perspective.

From an ecological perspective the Tumbler Ridge study is unique in a number of respects. First, an overall ecological paradigm underlies not only the design of the research but also the planning and development of the community. The planners used frameworks that can best be described as holistic, comprehensive, and evolutionary. Although the community (exosystem) is the focal system of this study, the key design elements being investigated will be examined at the microsystem, mesosystem, exosystem, and macrosystem levels. Second, a major source of information will be the perceptions of members of the community from a variety of role positions. Bronfenbrenner (1979) maintains that what matters for behavior and development is the environment as it is "perceived" rather than as it may exist in "objective" reality. Third, the study, methodologically speaking, is intended to be a creative combination of multiple data sources that will provide a description of the historical development and present state of the community and triangulate information on a set of key design elements. Fourth, a particularly important focus of the study is the interconnectedness of systems and system levels. The ecological approach emphasizes the importance of the connection between basic science and public policy and recommends that we move toward the functional integration of these two domains. An additional intent of the present study was to track public policy decisions at the macrosystem level and then examine the socio-behavioral consequences of these decisions at the microsystem, mesosystem, and exosystem levels.

BACKGROUND TO THE STUDY

In the fall of 1974, a multidisciplinary group of researchers from the University of Calgary, Alberta, were commissioned by the Privy Council Office in Ottawa to prepare a report for the National Design Council of Canada. They were asked to prepare "a report dealing with human aspirations, specifications for the social environment, and social guidelines for the physical environment in a frontier community in a hostile environment." More than 50 people participated in the task. On February 24, 1975, the final report entitled *Human Aspirations and Design Excellence*, was presented at the National Design Council Conference in Ottawa. Contrary to the principal investigator's skepticism regarding the future use of this report, it did emerge, as least initially, as a pivotal resource in the planning of Tumbler Ridge.

This previous investigation led to a new way of incorporating human aspirations into design processes. First, three major interdependent systems in our society were identified: *human systems*, *economic systems*, and *built-form systems*. It is clear that the designers of human, economic, and built-form systems are struggling with complex issues related to technological progress and its impact on the quality of human life-styles. Second, social science research was reviewed in an attempt to identify consistencies in human behavior that did not appear to be culture bound. Frontier communities were visited and residents of these communities were interviewed. An extensive review of the literature on human needs and aspirations was also completed. Out of these diverse experiences a list of three human aspirations that would be of use to designers was produced. These aspirations—security, diversity, and competency—were identified as three human commonalities that are defensible in terms of social science knowledge.

• *Security.* The word "security" is intended to include biological and psychological needs and wants. In the most elemental sense it should reaffirm and draw to our attention the fact that the human being is a biological animal whose survival is dependent on factors such as food, water, and clean air. In psychological terms, security can be used to connote the absence of fear and our need for affection.

• *Diversity.* One of the many characteristics humans share with other animals is an apparent "curiosity" drive. There is a great deal of recent research that suggests that people require a certain amount of change and stimulation. Experiments in "sensory deprivation" suggest that if someone is deprived of the opportunity to see, hear, smell, and touch, that person very quickly starts to exhibit signs of severe mental illness. However, the converse is also true. Too much stimulation or change is reacted to in a negative manner, and people tend to avoid situations that are too different from their past experiences. The need for diverse experiences is apparently consistent across cultures, but it varies with life-cycle stages.

• *Competency.* Some of the academic words that suggest elements of what we have in mind for "competency" are mastery of the environment, acquisition and exercise of skills, control, power, influence, and so forth. While the opportunity to acquire skills may diminish somewhat in later life, the opportunity to exercise one's skills seems to be important throughout life. According to some social scientists, the individual's concept of worth and value (self-concept) are closely connected to the opportunity for the individual to grow and develop, to exercise skills, to deal effectively with the environment, and to influence and change the present structure. Paternalism reduces our feelings of competency. Opportunities to participate in decision making and to act autonomously are likely to increase our feelings of competence.

Working definitions of each of these aspirations were then applied to each of the three systems—human, economic, and built-form—to generate lists of frontier community guidelines. In turn, these community guidelines were applied to the fairly extensive scenario of realities provided by the Design Council, resulting in a partial sketch of a frontier community that would satisfy human aspirations. While this investigation ended in recommendations for a fictional frontier community, the design and development of Tumbler Ridge represents an opportunity to document the consequences of implementing such recommendations.

THE TUMBLER RIDGE PROJECT

Since 1976 the Province of British Columbia, through its Ministry of Municipal Affairs and Housing, has engaged in settlement planning related to the exploration of coal resources in the northeastern part of the province.

> The Province's main objective is the implementation of its economic development policies including full employment, increased real income through resource allocation, price stability, and regional balance. The Province may implement its objectives by fostering and managing the growth of settlements associated with resource development. As the primary agency for settlement planning, the Ministry of Municipal Affairs and Housing is interested in developing, as soon as possible, a politically functioning, financially viable, well planned community with a relatively high level of services. It plays the key role in determining the organizational form and structure of the community, its level of services, its financial foundations, and, by its commitment to planning, the quality of the community. The Ministry wishes to pursue these objectives while minimizing its risks. (Paget & Rabnett, 1981, p. 156)

Tumbler Ridge, a completely new resource community, was chosen to be the first focus for the implementation of this policy decision. This community, which had about 1200 residents at the end of 1982, five years later had a population of 4300. It is located approximately 660 km north of Vancouver and 100 km southeast of its closest urban center, Dawson Creek. The major objective of the provincial government in the planning of this community was to "create a socially cohesive, financially viable, self-governing community, conducive to attracting and retaining a stable work force." A secondary, but perhaps more important objective, was to test a range of ideas that could be used in future resource community planning. In their early work on the design of the community, the planners identified three comprehensive themes:

1. A *natural theme*, stressing retention and management of the natural vegetation, use of local building materials, development of a sense of self-sufficiency and self-reliance, pursuit of outdoor leisure activities, and provision of rural lot development opportunities.
2. A *learning theme*, stressing continuing education; enrichment learning programs; individual motivation; community

initiated, developed, and managed programs; and experimental programs for learning disabilities.

3. An *experimental theme*, stressing the initiation, development, and monitoring of ongoing research projects such as wind and solar energy, land reclamation, building technology, and environmental management.

The Ministry of Economic Development, with overall leadership responsibility for the project, initiated and implemented comprehensive economic, transportation, environmental, manpower, and settlement planning. Under this overall economic planning umbrella, Municipal Affairs assumed lead responsibility for settlement planning. The Tumbler Ridge community plan provides a comprehensive development framework including a Social Plan, a Physical Plan, a Financial Plan, and an Organizational Plan (Thompson, Berwick, Pratt, & Partners, 1978).

A tremendous amount of time, money, and effort have been invested in the planning of Tumbler Ridge. Personnel from a number of government ministries, consultants, and residents of existing resource communities have provided input for the project. Numerous documents, monographs, and working papers have been developed. Throughout this process there has been a commitment to dialogue and participation. This was evident to a commendable degree throughout the planning phase.

Although there was a plan to document the history of the community and use surveys and other mechanisms to obtain resident input, no person or group was given the responsibility for overall evaluation of the project. In my view, the need to evaluate the development of Tumbler Ridge is absolutely crucial. It is a bold new experiment, which will produce information relevant and valuable to governments, resource companies, resource community residents, academics, and practitioners who are interested in social development. Based on this belief, a research proposal was submitted to the Social Sciences and Humanities Research Council of Canada. In April 1986 a research grant was awarded and work commenced in June 1986. The co-investigator for the research project is Dr. Alison Gill of Simon Fraser University.

Objectives of the Research

The major objectives of the research project are as follows:

1. to complete an inventory of existing information on the social and physical development of Tumbler Ridge;

2. to compile a detailed history of the first four years of the development of the community (1982–86);
3. to develop an accurate profile of the community as it presently exists;
4. to identify the process and degree of implementation of a number of key design elements in the development of the community;
5. to investigate the social and behavioral consequences of those design elements;
6. to interface the findings of this study with the existing social science research literature on new resource communities.

Theoretical Framework

The research was conceptualized within the general community planning framework that has been developed to guide planners in British Columbia (Paget & Rabnett, 1983). Resource community planning is primarily a process of reconciling the diverse interests of different actors in resource development—resource companies, workers, residents, and governments—each of which is influenced by different economic factors. The approach incorporates a style and method of decision making designed to deal with the uncertainty and economic volatility that are characteristic of the resource sector. The objective is to create resilient communities that are adaptable to changing conditions. The framework is not a definitive policy but is characterized as a "normative, process-oriented, analytical approach to decision-making" (Paget & Walisser, 1984, p. 110). This approach is analytical in that research and analysis are relied on as a basis for decision making. It is process-oriented because responsiveness to change is constantly sought, and it is normative in the sense that preferences are built into the process at critical points. The key notion embodied in the planning framework has been termed "disciplined incrementalism" (Paget & Rabnett, 1983), that is, decision making that combines features of disjointed/incremental and rational/comprehensive planning (Faludi, 1973).

The theoretical framework of the proposed study can be viewed, more specifically, in the context of the planning models used to direct the planning of Tumbler Ridge. These models, which were derived within the planning framework approach described above, are evolutionary in nature, changing as the community develops from the planning stage toward "maturity." From an ecological perspective, one can argue that this approach to planning is valid since it is responsive to the nature of the context, which undergoes a series of transitions. Characteristics of the three models are summarized in Figure 9.1.

Socially sensitive planning is the initial model. Social needs are of paramount concern, and planners' decisions are guided by findings of social science research. As shown in Figure 9.2 this approach integrates the theory and methods of social impact assessment (SIA) into the planning process, and thus the study can also be conceptualized within this broader theoretical framework of SIA. This approach is particularly strong in its ability to describe and assess the interconnectedness of the various systems involved.

During the implementation stage (see Figure 9.1) the focus is on translating social objectives into practice by means of "planning by invitation," which involves innovative organizational management and interagency coordination. This model evolves during the final planning phase into a model of community development based on participative community action, or "bottom-up planning."

Within the theoretical framework described above, the study represents the evaluative stage, an essential, but often neglected, phase of planning and SIA (Bowles, 1981; Lichfield et al., 1975). Ideally, evaluation should be a continual process, providing feedback into the planning process and offering direction to future new

FIGURE 9.1 Three Models of Planning

	PLANNING Phase	IMPLEMENTATION Phase	DEVELOPMENT Phase
	SOCIALLY SENSITIVE PLANNING	PLANNING BY INVITATION	COMMUNITY DEVELOPMENT
Client	province	local government	community
Style	top-down technical directive	horizontal invitational/ inter-organizational creative management	bottom-up community-based participative
Focus	product plans budgets negotiations	process innovative structural design institutional development coordination	community action
Use of Information	factual processed strategic definitive social	instrumental organizational policy relevant experimental	normative personal issue focussed speculative spatial

SOURCE: Paget and Rabnett, 1983, p. 28.

FIGURE 9.2   Socially Sensitive Resource Community Planning

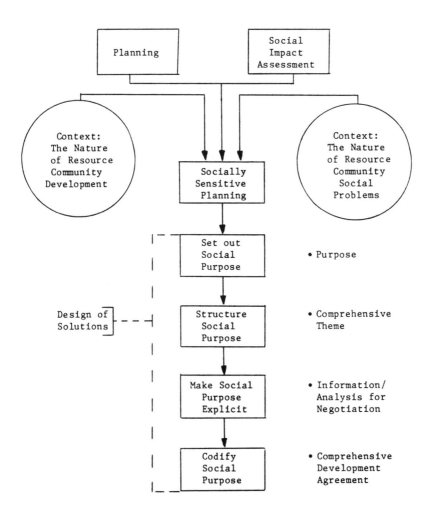

SOURCE: Paget and Rabnett, 1981b, p. 150.   Used by permission.

town development (Shera, 1984). In response to this need, the implementation and consequences of key social principles that guided the planning of Tumbler Ridge were investigated.

These key social principles were originally derived from social science research and address three basic objectives in resource town planning: social cohesion, reduced labor turnover, and residential satisfaction. To meet these objectives in Tumbler Ridge, the planners used a set of social principles (described in Figure 9.3) to guide the design of various elements of the community. Assumptions were made about social and behavioral responses to these design elements. The selected social principles, design elements, and underlying social and behavioral assumptions that provided the focus of the study are outlined in Figure 9.4.

Virtually all of the design elements illustrate how the planners tried, at least implicitly, to use an ecological approach to the development and implementation of the community. Physical design features such as the town center, neighborhoods, and crime prevention demonstrate an awareness of the impact of physical factors on human behavior. Another major feature is the use of the town hall, the health and social service center, and the community recreation center as physical and psychological "anchors" for the community. Special provisions for teens and women, a social development officer, and an emphasis on citizen participation reinforce the ecological concepts of reciprocity and interconnectedness. An important point to underscore is the utility of the ecological paradigm in the design and creation of human environments.

From a research perspective, the present study, at the microsystem level, investigated individuals' perceptions and behavior in the home, work, and community environments. At the mesosystem level, we are particularly interested in the connections among the home, work, and community environments. A major issue in this respect is "shift work," which has a variety of impacts on individuals' and families' behavior both within and outside the community. In terms of exosystem, we are paying particular attention to the concept of neighborhood, the gradual development of local government, the quality of commercial services available, and the dynamics of the commuting labor force. As previously mentioned, from a macrosystem perspective, the study will address such central issues as planning models, local government approaches to development, resource development in general, and implications for other similar settings. To provide a comparative base, the Tumbler Ridge data will be compared with the results of other studies on resource communities.

FIGURE 9.3   Social Principles for Development

OVERALL   To create a socially cohesive, financially viable,
OBJECTIVE   self-governing community conducive to attracting
and retaining a stable workforce

Choice: It is important that residents locate in Tumbler Ridge by choice and not just by economic circumstances. Taking personal responsibility for this decision encourages a commitment to the town.

Commitment: A successful community requires a commitment to plan, a commitment to participate, and a commitment to produce a community that fosters stability, a sense of belonging, and personal growth.

Challenge: Each participant has a responsibility to foster a new way of thinking that emphasizes the challenge of the frontier and appeals to the pioneer spirit.

Self-reliance: The focus should be on the individual and what he or she can do for himself or herself.

Participation: The development of the town must reflect the needs of the people who live there if it is to become a cohesive, stable community.

Integration: Integration of social and health facilities and programs makes sense both from a service delivery and a cost-effectiveness perspective.

Equity: Tumbler Ridge must plan for those people who experience the greatest difficulty in new towns.

Fiscal Responsibility: The town will one day stand on its own and be financially responsible. The town, therefore, must be designed to be financially viable.

Environmental Sensitivity: It is to the advantage of the community and the residents to develop as harmonious a relationship as possible between the town and its environment.

Flexibility: All decisions, policies, and programs must be flexible to accommodate unexpected changes in the development of the town.

Source: Paget and Rabnett, 1981a, p.9.

FIGURE 9.4   Assumed Social-Behavioral Consequences of Design Elements
Derived from Selected Social Principles

PRINCIPLE/Design Element	Expected Consequence
1. CHOICE	
Selective hiring	Commitment to community will be increased if residents choose to move to Tumbler Ridge for other than economic reasons.
Housing options	Residents will experience less stress and financial burden if housing options are within the limits of their financial constraints.
Limited range of good quality services	Increased resident satisfaction, and commitment to remain in the community.
2. COMMITMENT	
Local government at early stage	Increased commitment by community residents.
Home ownership	Increased residential stability
3. CHALLENGE	
Base-level service provision	Appeals to pioneer spirit and encourages involvement and mutual aid.
Network of open spaces	Encourages exploration and interaction.
4. SELF-RELIANCE	
Local government	Encourages resident responsibility and involvement.
Reduced dependence on employer and government social services	Contributes to social cohesion through participation and mutual aid activities in the community.
Base facility recreation programs	Community-generated facilities will reduce financial and psychological dependence on government and mining companies and encourage social cohesion.
5. PARTICIPATION	
Social development officer	Acceleration of social integration.
Local government	Increased resident involvement and commitment.
Inhabitants involved in modifying and adapting community	Increased satisfaction with the community, social cohesion and commitment to remain.
Crime Prevention through Environmental Design (CPTED)	Residents' involvement in surveillance to prevent vandalism increases residents' commitment and satisfaction.

FIGURE 9.4   *continued*

6. INTEGRATION

Integration of social   Provides focus for social development
and health facilities   and increases resident satisfaction
                                with the community

Neighborhood design   Encourages social interaction and
                                cohesion.

One town center   Provides sense of stability and perman-
                                ence and encourages social interaction.

7. EQUITY

Social provisions   Increases community satisfaction of
for women   women, who have greater difficulty than
                                men in developing social ties in re-
                                source communities.

Teen lounge   Teen-agers will utilize lounge and not
                                "hang out" in areas not intended for
                                that purpose.

8. ENVIRONMENTAL SENSITIVITY

Design and orientation   Improves quality of life and increases
of streets and buildings   residents' satisfaction with the com-
with sensitivity to wind,   munity.
sun, and snow

Design of town hall as a   Provides sense of place and creates
community focal point   community identity.

SCHOLARLY AND SOCIAL RELEVANCE

There is an extensive interdisciplinary body of literature relevant to the study, which falls into the following categories: (1) resource community planning, (2) social and behavioral aspects of design, and (3) social impact assessment. A brief overview of research in each of these areas is presented, with a focus on aspects especially pertinent to the study.

Resource Community Planning

New resource communities experience a wide range of problems, which fall into two major types: problems of the town and problems of implementation. Problems related to the town include such things as the level of employment stability of the population, the quality of services available, availability of housing, and afford-ability. Problems related to implementation, or the financing and delivery of the community, include such things as poor planning, inadequate growth, inequitable risk, timing of services, and rapid growth. It is extremely difficult to separate social problems in the town from problems of delivering and financing the town—they are interdependent and hence few problems lack a social dimension.

The social consequences of past approaches are well docu-mented: Glick and Glick (1981); Lucas (1971); and Riffel (1975) pro-vide good overall discussions of the social planning requirements of resource-based communities. More recently the development of the concept of social impact assessment has focussed attention on the social implications of large-scale projects (Tester & Meikes, 1981), the relationships between SIA and policy science (Torgerson, 1980), and the relationship between SIA and community planning (Paget & Rabnett, 1981a). It is the latter issue that is of primary concern. While SIA and community planning are fundamentally different activities, both are rooted in a basic social ethic and a belief that quality of in-formation should lead to better decisions (Paget & Rabnett, 1981b).

In recent research conducted by Roberts and Fisher (1984, p. 163) in Fort McMurray, Leaf Rapids, Kitimat, Sparwood, and Elkford, community residents provided a range of recommendations for planners of new resource communities.

1. anticipatory rather than reactive planning;
2. appropriate, affordable housing designed for the local win-ter and snow conditions;
3. resident involvement in the planning process at the com-munity and housing levels;
4. planners and consultants who have spent time in the com-munity and in the region so they are aware of the problems;
5. no high density housing;
6. draw on experience of other communities/situations/ residents to plan a new community;
7. have health, education, and commercial services available in the community before residents arrive.

This research not only supported recommendations from pre-vious studies but also reflected many aspects of the approach being used in the development of Tumbler Ridge.

Social and Behavioral Aspects of Design

During the last two decades, analysis of the relationship between human behavior and the physical environment has attracted researchers from many fields in the social sciences, such as psychology, sociology, geography, and anthropology, and from the environmental design fields of architecture, urban and regional planning, landscape architecture, and interior design. Thus, environmental design research ranges from micro-scale considerations of personal space (Sommer, 1969) to macro-scale studies at the urban or regional level (Gould & White, 1974; Lynch, 1960). Overviews of the field include those by Proshansky and associates (1970), Saarinen (1976), and Holohan (1982). The common focus of environmental design studies is the attempt to understand human cognition and behavior as the central feature in design decisions. Common methodological procedures, primarily derived and adapted from those in psychology, unite this interdisciplinary field (Michelson, 1975; Zeisel, 1981).

Translation of social and behavioral findings into design features is often difficult. However, planners increasingly are attempting to incorporate this approach into the planning framework. In Canada, for example, at the federal level of government, environmental design research techniques are being applied to evaluate design and building performance in a variety of settings, including health care facilities, office buildings, airports, national parks, and residential environments (Harvey & Vischer, 1984). Of particular significance to the proposed research is the work of architect Christopher Alexander and associates, especially the "pattern language" approach (Alexander, Ishikawa, & Silverstein, 1977). Alexander's philosophical approach to design, which embodies environmental design principles for creating a "sense of place," has played a significant role in guiding design in Tumbler Ridge. Other research in the field that has influenced design in Tumbler Ridge includes that of Clare Cooper (1975) on Easter Hill Village and Oscar Newman (1972) on "defensible space," and Herbert Gans' (1962) work on urban social needs.

Social Impact Assessment

Social impact assessment (SIA) has been defined as assessing the effects of projects or policies on people—the human implications of what we do. It attempts to forecast the effect that a proposed development will have on the quality of life—the traditions, life-styles, interpersonal relations, institutions and living environment. (D'Amore & Rittenberg, 1978, p. 9)

A rapidly growing body of literature addresses these issues within the conceptual and methodological framework of SIA. Overviews of work in the field include those by Wolf (1974), Finsterbusch and Wolf (1977), Glickfield and associates (1978), and Carley and Derow (1980). More specifically, Bowles (1981) reviews SIA research relating to small Canadian communities in the resource hinterland, with an emphasis on aspects of social vitality of the communities, viability of the local economy, and internal political efficacy.

Conceptualization of SIA and the associated methodological procedures are fairly well documented (Finsterbusch & Wolf, 1977). In SIA, social indicators are utilized to permit judgments about how the quality of life is changed by a new project (Olsen & Merwin, 1977). While the majority of SIA research is predictive, focussing on the pre-impact stage, Johnson and Burdge (1974) stress the need for engaging in post-impact studies. In the context of new towns, Keller's (1978) study of Twin Rivers, New Jersey is one of the few relevant examples.

In Canada, SIA at the federal level has become a component of the activities of the Federal Environmental Assessment and Review Office (FEARO). This has, since the 1970s, generated a considerable Canadian contribution to SIA research at both a conceptual and methodological level (Blishen et. al., 1979; Boothroyd, 1975; Di Santo et al., 1978) and an applied level (Kupfer & Hobart, 1978; Stager, 1974).

METHODOLOGY

Given the scope of the present investigation within an ecological framework, a multiple-method, multiperspective approach was employed (Ball, 1977). The advantages of employing a "systemic perspectivism" approach in research have been articulated as follows:

> While orthodox definitions of "objectivity" lead to a preoccupation with data reliability, by which is usually meant the achievement of increased internal consistency from a particular point of view, the principle of perspectivism demands that the investigative sociologist strive for increased validity through the pooling of observations gained from different vantage points. Both discov-

ery and verification would proceed by a strategy of pinpointing data through a triangulation of methodologies (Webb, 1970), with the difference being that the triangulation would operate on several "planes" of social reality. (Ball, 1977, p. 12)

In recent years the use of naturalistic paradigms, multiple methods, and combinations of quantitative and qualitative data has gained credibility (Lincoln & Guba, 1985). These paradigms stress a holistic approach, the importance of perspectivism, and an accent on the study of relationships and processes rather than entities and structures. In addition to the overall paradigmatic approach to the study, it is crucial to identify some of the procedural dimensions to be considered. In many respects one cannot separate the "process" of the research from the results. In this regard two models are instructive. The first, described as the illuminative model, consists of three stages: observation and involvement, further inquiry, and explanation. This approach employs multiple sources of information, progressive focussing, and triangulation. It is not, however, a standard methodological package, but instead is a research strategy that aims to be both adaptable and eclectic in addressing the problem at hand. The use of interpretive human insight and skill is encouraged rather than discouraged, and research workers in this area need not only technical and intellectual capability but also interpersonal skills.

To carry out successful research in a social system as large as a community one must proceed carefully. This is where some of the literature and models of consultation can be helpful. A particularly useful one, vis-à-vis research and evaluation, has been presented by Jacobsen (1974). His model consists of six stages; translated into terminology relevant to the present investigation, it is as follows: (1) entry into the community; (2) relationship building; (3) community assessment; (4) finalization of data collection strategies/agreements; (5) data collection; and (6) analysis, presentation, and termination. The procedural approach employed in this investigation was a blending of those two models. The overall data collection strategy for the research project consisted of three major stages of research activity.

Stage I: Inventory of Existing Information

The purpose of this stage was to prepare a catalog of available information on Tumbler Ridge, with specific reference to the implementation and socio-behavioral consequences of the design elements being investigated. The types of information that were cataloged included:

1. planning reports/working papers;
2. papers presented at conferences/published articles;
3. newspaper/magazine articles;
4. audio-visual reports;
5. minutes/reports of the commissioner and local council;
6. reports/newsletters of human services organizations and other local associations in Tumbler Ridge;
7. focussed literature review on the social dimensions of re-source communities.

Stage II: Documentation of Development Processes

The purpose of this stage was to develop a historical profile of the development of the community, with particular emphasis on its social development. This included the completion of:

1. a profile of the development and present nature of the com-munity;
2. compilation of a set of community indicators, including population, economic growth, employment (men/women/youth), business success/failure, criminal activity, labor turnover, ethnic composition, spatial distribution, and com-munity participation (municipal and school board elections, etc.);
3. interviews with key informants (Ministry personnel and consultants) responsible for the social planning conducted prior to the building of the community;
4. interviews with key informants (local) responsible for the implementation of the social aspects of the plan and the community development processes employed to facilitate social development objectives (Town Commissioner, mem-bers of the Community Association and Municipal Council, Social Development officers, company liaison persons, etc.).

Interviews were used to obtain information from planners and architects. Information obtained from these respondents was related to the underlying reasons for design decisions and to a reconstruc-tion of relevant events in the planning process. Interviews were also used to elicit information, opinions, and attitudes from key informants within the community and from the two coal mining companies, Quintette Coal and Bullmoose. A primary objective of these discussions was to reconstruct the basis of decision making in the evolution of the community as it relates to the design elements being examined. Opinions and attitudes of these informants provided additional perspectives to those elicited from residents.

Stage III: Primary Data Collection

The purpose of this stage was to gather data from a range of community residents and groups on such issues as reason for coming to Tumbler Ridge, participation in the community, use of community services, comparisons with other similar mining towns, and perceptions regarding housing, social services, recreation, community spirit, etc. The issues addressed in these data collection methods were derived primarily from the community design elements being investigated. The components of this stage of the research project included:

1. Household survey of residents
2. Survey of secondary school children
3. Survey of elementary school children
4. Interviews with groups of key informants from the following areas:
     a) Health and social services
     b) Recreational organizations/associations
     c) Local government
     d) Business sector
     e) Education
     f) Long-term residents
5. Mail survey of commuters.

Information from residents was the most extensive data source, and various techniques were employed, including household and school surveys, group interviews, and a commuter questionnaire. A total of 270 household surveys, or 86 percent of the sample, was completed. The questionnaires used in the surveys were designed to obtain information regarding the key design elements and their social and behavioral consequences.

A modified form of the household questionnaire was used in the survey of all secondary school children, and a one-page questionnaire was given to all elementary school children. In our view these components of the study are extremely important because very few studies of this nature have obtained the views of children in such communities. We feel this is a much needed perspective on the development of new resource communities.

The four major component of this stage of the research was the collection of information from various "constituent" groups in the community. The groups identified represent a broad cross-section of the community. The group method employed in this study is a very effective and efficient manner of data collection. The principal investigator used this technique successfully in previous studies (Shera & Turner, 1981). Lists of constituent-group members are generated and then, depending on the size of the list, either the entire group or a random sample is selected to participate in a meeting of one-and-a-half to two hours in length. Group size is limited to 15 participants, who complete a questionnaire for the first 20 minutes of the meeting and are then asked by a group facilitator to discuss issues of central concern to the study. These discussions are tape-recorded, and highlights are put on flip chart paper as the group proceeds with the discussion. The facilitator ensures that all topics have been covered and allows participants a final 10 minutes to add any further comments to their questionnaire. As previously mentioned, this process not only gathers information in an effective and efficient manner, but it also uses the discussion and exchange to reflect on and in most cases clarify and crystallize various perspectives on the issues under investigation.

The final component, the mailed questionnaire to commuters, was initially an unintended part of the study. As the research proceeded, it became clear that commuters were a significant dynamic in the life of the community. Although there were various popular ideas about why workers commuted, no hard evidence was available. Almost 40 percent of the commuters completed questionnaires.

The data base that has been assembled is enormous. A few highlights of the results are presented below to illustrate the utility of the ecological approach.

We asked residents, secondary school students, and key informants to rate community services. The overall profiles of ratings for each of the groups are remarkably similar. In general, the schools, day care facilities, programs/activities for youth, police, library, service and recreation clubs, and municipal services are seen as being above average. The two areas that received poor ratings were clothing/furniture/retail stores and entertainment. Residents also felt that the grocery stores were below average.

Another useful method of assessing community facilities and services is through regional comparisons. Residents and key informants were asked to compare Tumbler Ridge with neighboring communities on a variety of factors. Again both groups were very similar in their profile of ratings. Those aspects of Tumbler Ridge that were seen as being worse than other communities included grocery prices, choice of retail goods, housing costs, housing choice, health and social services, and overall cost of living. Those which were seen as better included opportunity for community involvement, attractiveness, and Tumbler Ridge as a place to raise children.

The results, then, on community services and regional comparisons converge and generally indicate that various groups of residents view these factors in a very similar manner. On the other hand, the detailed results on community satisfaction tell a somewhat different story. When asked to rate Tumbler Ridge on a variety of dimensions, residents and secondary school students provided very similar ratings. The key informants who had been involved in the group meetings, however, rated Tumbler Ridge higher on opportunity to participate, friendliness, challenge, a safe place to hire, attractiveness, a place to get ahead, and ability of residents to influence development. From an ecological perspective, these results make perfectly good sense. Key informants are, by definition, more involved and committed to the community and their various roles as leaders in the community affect their perceptions of the setting.

A further set of data on community satisfaction will reinforce this point. Residents, group participants, and elementary and secondary students were asked to rate Tumbler Ridge as a place to live. A five-point scale was used with five being very good and one, very poor. The following mean ratings were obtained from each group: residents, 3.37; secondary school students, 3.38; group participants, 3.84; and elementary school students, 4.00. Again we can use an ecological perspective and the notion of developmental contexts to interpret these data. The resident and secondary school student ratings are essentially the same. Relevant literature would suggest that socialization within the family context would produce similar attitudes toward the outside world. There will, however, also be differences in attitudes and values, but in the case of the community as the "context," one would expect similar ratings. The group participants or key informants provided a mean rating of 3.84, and this reinforces the previously discussed notion that roles influence perceptions. The elementary school students provide a very positive rating of the community (4.00), which one can explain by understanding the developmental contexts in which the children are embedded. The elementary schools in Tumbler Ridge are attractive, well equipped, and have enthusiastic and committed teachers. Tumbler Ridge is very much a child-oriented town. The birth rate is one of the highest in the province; there are many young families in the town; and the Community Centre gears many of its programs to young children. Given these positive, child-oriented developmental contexts, one can understand why children rate the community as highly as they do. These findings strongly reinforce the importance of obtaining views from different segments of the population when conducting community studies. They also highlight the importance of background and contextual information to the interpretation of quantitative data.

CONCLUSION

As previously mentioned, we are just in the early stages of data analysis. The ecological perspective has been extremely useful in both the design of the study and in the interpretation of the results. The completed research reports and forthcoming articles on Tumbler Ridge should make a significant contribution to the existing research literature on resource community planning, social and behavioral aspects of design, and social impact assessment. Given the interdisciplinary nature of the study it is also anticipated that contributions will be made to other areas of research. The multi-method, interdisciplinary approach used in the execution of the research may also provide some insights to other researchers. At the applied level, the research should provide useful planning and policy information for federal and provincial government officials, coal companies, other resource communities, and others involved in the design and implementation of new resource communities.

REFERENCES

Alexander, C.; Ishikawa, S.; & Silverstein, M. (1977). *A pattern language.* New York: Oxford University Press.

Ball, R. (1977). Equitable evaluation through investigative sociology. *Sociological Focus 10,*(1), 1–14.

Blishen, B. R.; Lockhart, A.; Craib, P.; & Lockhart, E. (1979). *Socio-economic impact model for northern development.* Ottawa: Department of Indian Affairs and Northern Development.

Boothroyd, P. (1975). *Review of the state of the art of social impact research in Canada.* Ottawa: Ministry of State for Urban Affairs.

Bowles, R. T. (1981). *Social impact assessment in small communities.* Toronto: Butterworth & Co. (Canada) Ltd.

Bronfenbrenner, U. (1979). *The ecology of human development: Experiments by nature and design.* Cambridge, MA: Harvard University Press.

Carley, M. J. & Derow, E. O. (1980). *Social impact assessment: A cross-disciplinary guide to the literature.* London: Policy Studies Institute.

Cooper, C. (1975). *Easter Hill Village.* New York: Free Press.

D'Amore, L. J. & Rittenberg, S. (1978). Social impact assessment: A state of the art review. *Urban Forum, 3,*(6), 3–21.

Di Santo, J.; Frideres, J.; Fleising, U.; & Goldenberg, S. (1978). Industry, government and community relations in social impact assessment. Paper presented to the First Canadian Symposium on Social Impact Assessment, Banff, Alberta, November 30–December 2, 1978.

Faludi, A. (Ed.) (1973). *A reader in planning theory.* Oxford: Pergamon Press.

Finsterbusch, K. & Wolf, C. P. (Eds.) (1977). *Methodology of social impact assessment.* Stroudsburg, PA: Dowden, Hutchinson & Ross.

Gans, H. (1962). *The urban villagers.* New York: Free Press.

Glick, I. & Glick, M. (1981). *Boom towns: A quest for well-being.* Edmonton: Canadian Mental Health Association.

Glickfield, M.; Whitney, T.; & Grigsby, J. E. (1978). *A selective analytical bibliography for social impact assessment.* Monticello, IL: Council of Planning Librarians (Exchange Bibliography No. 1562).

Gould, P. R. & White, R. R. (1974). *Mental maps.* Harmondsworth, Middlesex: Penguin.

Harvey, J. & Vischer, J. (1984). Environmental design research in Canada: Innovative government intervention. Workshop summary in D. Duerk & D. Campbell (Eds.), *The challenge of diversity.* Proceedings of Environmental Design Research Association Conference 15:278. Washington, D.C.: EDRA.

Hawkes, F. J. & Associates (1975). *Human aspirations and design excellence.* Calgary: The School of Social Welfare, University of Calgary.

Holohan, C. J. (1982). *Environmental psychology.* New York: Random House.

Jacobsen, E. A. (1974). *Program evaluation and psychological consultation.* Unpublished monograph. University of Michigan, Community Mental Health Program.

Johnson, S. & Burdge, R. (1974). Social impact statements: A tentative methodology. In C. P. Wolf (Ed.), *Social impact assessment.* Section 2 of Part I in D. C. Carson (Ed.), *Man–environment interactions* (pp. 69–84). Stroudsburg, PA: Dowden, Hutchinson and Ross.

Keller, S. (1978). Design and the quality of life in a new community. In J. M. Yinger & S. J. Cutler (Eds.), *Major social issues: An interdisciplinary view.* New York: Free Press.

Kupfer, G. & Hobart, C. (1978). Impact of oil exploration work on an Inuit community. *Arctic Anthropology, XV*(1), 58–67.

Lichfield, N.; Kettle, P.; & Whitbread, M. (1975). *Evaluation in the planning process.* Oxford: Pergamon Press.

Lincoln, Y. & Guba, E. (1985). *Naturalistic inquiry.* Beverly Hills, CA: Sage Publications.

Lucas, R. (1971). *Minetown, milltown, railtown: Life in Canadian communities of single industry.* Toronto: University of Toronto Press.

Lynch, K. (1960). *The image of the city.* Cambridge, MA: M.I.T. Press.

Michelson, W. (Ed.) (1975). *Behavioral research methods in environmental design.* Stroudsburg, PA: Dowden, Hutchinson and Ross.

Newman, O. (1972). *Defensible space.* New York: Macmillan.

Olsen, M. E. & Merwin, D. J. (1977). Towards a methodology for conducting social impact assessments using quality of social life indicators. In K. Finsterbusch & C. P. Wolf (Eds.), *Methodology of social impact assessment* (pp. 43–63). Stroudsburg, PA: Dowden, Hutchinson and Ross.

Paget, G. & Rabnett, R. (1981a). Planning for large scale developments: Responding creatively with a sensitivity to social needs. Victoria: Policy and Research Branch, Ministry of Municipal Affairs.

Paget, G. & Rabnett, R. (1981b). Socially responsive community planning: Applied social impact assessment. In F. Tester & W. Mykes (Eds.), *Social impact assessment: Theory, method and practice.* Calgary: Kananaskis Centre for Environmental Research, University of Calgary.

Paget, G. & Rabnett, R. (1983). *The need for changing models of planning: Developing resource based communities.* U. B. C. Planning Papers, Canadian Planning Issues, No. 6. Vancouver: School of Community and Regional Planning, U. B. C.

Paget, G. & Walisser, B. (1984). The development of mining communities in British Columbia: Resilience through local government. In *Mining communities: Hard lessons for the future.* Proceedings of the Twelfth Policy Discussion Seminar, Centre for Resource Studies, Kingston, Ontario, September 27–29, 1983, 96–150.

Parlett, M. & Hamilton, D. (1976). Evaluation as illumination: A new approach to the study of innovatory programs. *Evaluation Studies Review Annual.* Gene V. Glass (Ed.), Volume 1. Beverly Hills, CA: Sage Publications.

Proshansky, H.; Ittelson, W.; & Rivlin, L. (Eds.) (1970). *Environmental psychology: Man and his physical setting.* New York: Holt, Rinehart & Winston.

Riffel, J. A. (1975). *Quality of life in resource towns.* Ottawa: Ministry of State for Urban Affairs.

Roberts, R. & Fisher, J. (1984). Canadian resource communities: The resident's perspective in the 1980's. In *Mining communities: Hard lessons for the future.* Proceedings of the Twelfth Policy Discussion Seminar, Centre for Resource Studies, Kingston, Ontario, September 27–29, 1983, 151–70.

Saarinen, T. (1976). *Environmental planning: Perception and behavior.* Boston: Houghton Mifflin.

Shera, W. (1984). New resource communities: Dilemmas in social development. In T. Walz & M. Jacobsen (Eds.), *Social development issues: Alternative approaches to meeting human needs, 8*(1,2), 144–57.

Shera, W. & Turner, D. (1981). *The house arrest program: An evaluation.* Victoria: School of Social Work, University of Victoria.

Sommer, R. (1969). *Personal space: A behavioral basis for design.* Englewood Cliffs, NJ: Prentice-Hall.

Stager, J. (1974). *Old crow, Y.T. and the proposed northern gas pipeline.* Environmental-Social Committee, Northern Pipelines, Task Force on Northern Oil Development, Report No. 74–21, Ottawa: Information Canada.

Tester, F. & Mykes, W. (1981). *Social impact assessment: Theory, method and practice.* Calgary: Kananaskis Centre for Environmental Research, University of Calgary.

Thompson, Berwick, Pratt, & Partners (1978). *Conceptual plan: Tumbler Ridge, northeast sector, B.C., a physical plan, social plan, financial plan, organizational plan.* For Ministry of Municipal Affairs and Housing, Government of British Columbia.

Torgerson, D. (1980). *Industrialization and assessment: Social impact as a social phenomenon.* Toronto: York University Publications in Northern Studies.

Webb, E. J. (1970). Unconventionality, triangulation and inference. In N. K. Denzin (Ed.), *Sociological methods*, pp. 449–455. Chicago: Aldine-Atherton.

Wolf, C. P. (Ed.) (1974). *Social impact assessment.* Section 2 of Part I in D. C. Carson (Ed.), *Man–environment interactions.* Stroudsburg, PA: Dowden, Hutchinson and Ross.

Zeisel, J. (1981). *Inquiry by design: Tools for environment-behavior research.* Monterey, CA: Brooks/Cole.

10

IMPLEMENTING THE ECOLOGICAL PERSPECTIVE IN POLICY AND PRACTICE: PROBLEMS AND PROSPECTS

BRIAN WHARF

Brian Wharf, in this, the final chapter, considers the ecological perspective as it relates to both social policy and social welfare practice. Arguing that the ecological approach provides a conceptual model for the reform of both policy and practice in North American social welfare, Wharf explores the reasons why such reforms have not taken place and suggests strategies for their implementation.

The ecological perspective is rooted in the position that the health and well-being of individuals are intimately connected to and affected by the environment in which the individuals live. The distinguishing characteristic of the ecological perspective is its insistence on a holistic understanding of individuals within their social context and on tracing the connections between individuals and the environmental forces that affect them. Thus, ecological researchers have examined the consequences for families living in poverty (National Council of Welfare, 1975; Pelton, 1981), without support from families and friends (Pancoast, 1980; Whittaker, 1983), and in neighborhoods characterized by poor environmental conditions (Garbarino & Sherman, 1980; Warren, 1980). The literature resulting from the above and many other researchers is rich in detail and convincing in its support of the position that environments at all levels affect family life and influence the capacity of parents to care for children.

From this anchorpoint the perspective goes on to argue that it makes no sense to hold the poor, the abusive parent, and many other casualties of our society totally responsible for their plight when, in fact, environmental forces over which they have little control contribute significantly to their situations. Blaming the victim results in further frustrations not only for the individual but also for the family, the community, and ultimately society. At the same time, the ecological perspective demands that individuals be as responsible and as self-reliant as possible. As will be developed later in the chapter, the principal helping strategies for ecological practitioners are to identify and build strengths of individuals and families, empowering them to take charge of their situation and to change aspects of environments that hinder the development of healthy families.

There is not sufficient space in this chapter to trace all the connections between the ecological perspective and its historical antecedents. But it is important to note that the ecological perspective has an impeccable lineage in its direct connection to the work of C. Wright Mills in sociology, William Schwartz in social work, and Emory Cowen in community psychology, to say nothing of its roots in biology and systems theory. Some of these connections are described below.

Mills developed the classic distinction between personal troubles and public issues.

> Troubles occur within the character of the individual and within the range of his immediate relations with others; they have to do with self and those limited areas of social life of which he is directly and personally aware. Issues have to do with matters that transcend these local environments of the individual and the range of his inner life. An issue is a public matter. (Mills, 1959, p. 8)

The writings of William Schwartz are instructive for his careful review of the attempts of the social work profession to deal with both private troubles and public issues and for his imaginative but never implemented proposals for dealing with both. Schwartz argued that the consequence of compartmentalizing and dealing separately with personal troubles and public issues is

> a half licensed profession and a living symbol of the schizophrenia incurred by the failure to understand the connections between private troubles and public issues. To create a department for each would in fact institutionalize the very evils they mean to solve. The clinicians (dealing with private troubles) would be shielded from any further pressure to bring the weight of their

experience with people in trouble to bear on the formulation of public policy; and the social planners would be set free from the headaches of practice and left alone to fashion their expertise not from the struggles and sufferings of people but from their own clever and speculating minds. (Schwartz, 1969, p. 35)

For Schwartz the answer lay not in compartmentalizing, but in concentrating on the connection between troubles and issues. Social workers should focus attention on the interrelationships between people and their environment and seek to change these relationships. Schwartz' work has been developed by Allan Pincus and Anne Minahan (1973) in their conception of the generic approach to practice, and more recently by Carel Germain and Alex Gitterman (1980), who have located their conceptualizations within an ecological perspective. For all, connections between systems (individual, families, communities, and societies) are the key ingredient in understanding and changing the behavior of individuals and modifying larger systems.

A related body of thought and writing can be found in the work of community psychologists like Emory Cowen. Dissatisfied with services that simply respond to established problems, Cowen has struggled with the need to conceptualize a case for prevention in fields such as mental health. Recognizing that there is little to sustain an argument for primary prevention, Cowen has focussed on "baby steps" to primary prevention and has advanced the assumption that two such baby steps are enhancing the competence of clients and changing negative environmental conditions. Both of these strategies are "targetable on an impersonal basis to all people rather than to individuals already experiencing distress" (Cowen, 1977, p. 6).

A quotation from Bronfenbrenner will serve to make explicit the particular focus of this chapter.

The ecology of human development involves the scientific study of the progressive mutual adaptation between an active growing human being, and the changing properties of the immediate settings in which the developing person lives as this process is affected by relationships between these settings and by the larger contexts in which the settings are embedded. (Bronfenbrenner, 1979, p. 21)

In this chapter, "settings" are viewed and discussed as communities, and although social policies are far from synonymous with "larger contexts," they are typically seen as the instruments used by societies to improve social conditions and to protect against or offset the impact of economic, industrial, and other forces.

It is apparent from the above and from the writings of other ecological theorists that there are two essential characteristics of the ecological perspective. First, there are connections between individuals, communities, and the larger social forces, and second, not only do these change but so do the connections between them. The challenge is, therefore, to identify social policies and practices that promote change and in a direction that extends opportunity for individuals and enhances their well-being. It is the contention of this chapter that the present policies and practices in family and children's services are residual and crisis-oriented. They serve to confirm the present status of individuals, and they do not give attention to connections between the various units. In contrast, the universal policies and empowering practices outlined in this chapter do promote growth and change and are, therefore, consistent with an ecological perspective. A concern of this chapter is to make the case that the ecological perspective, which is usually applied to practice with individuals, families, and community, can be extended to embrace policies at the state and national levels.

It needs to be added that in federal countries like the United States and Canada, often there is not a close connection between national social policies, provincial or state legislation for families and children, and the actual practice methods used by workers in local agencies. Thus it is possible to have universal policies in place nationally but for a residual approach to prevail in family and children's services in local communities. The writer's home province of British Columbia serves as one example, whereas in many states in the United States the reverse of empowering practice and residual national policies exists. In federal countries, state or provincial governments occupy a crucial position in the provision of human services. They are uniquely placed to consider the impact of national policies on communities and to articulate the needs of communities to the national level. The distinctive role of state and provincial governments is addressed in the concluding portion of the chapter.

Before we turn to an examination of social policy and practice, a few words about the societal context are required. In both Canada and the United States social conditions are turbulent and troublesome. Unemployment rates are high, ranging in Canada from 14 percent in some regions to 8 percent in Ontario. But even the latter percentage would have been considered unacceptable some 20 years ago. Rates of poverty, of family breakdown, of teen-age pregnancies, of child abuse and neglect, of violence against women, and of teen-age depression and suicide are also high.

Edward Zigler provides this chilling description of family life in the United States.

> The divorce rate is still high; it has dipped slightly, but approximately 50% of all marriages will end in divorce. The teenage pregnancy problem in our nation remains a major tragedy. Teenagers account for about one million pregnancies per years, with 600,000 live births. Twenty percent of all children born in our society are born to unmarried mothers. In some hospitals in our poor urban areas, over 50% of the women giving birth are single teenagers. Another striking change is the number of single-parent families. Single-parent families is a euphemism; 90% are headed by women. Today, one in four of all children are being raised in single-parent homes. Such figures make clear why we have now entered the period characterized by what has been termed "the feminization of poverty." Children, single women, and the elderly now constitute the majority of poor people in the United States. Children comprise 25% of our nation's poor. (Zigler, 1986, p. 9)

In Canada the societal context is only marginally better. In Quebec, for example, the situation of families in the 1980s is as follows:

> 20 percent are headed by a lone parent, while 70 percent of families on welfare are lone-parent families; 48 percent of married women work for wages; the number of divorces rose 80 percent from 1969 to 1982; common law marriages represent 20 percent of total marriages, and 15.6 percent of births were "out of wedlock." (Guberman, 1987, p. 18)

The reasons for this state of affairs are a matter for fierce and unresolved debate. The reasons range from the loss of religion as a central force, to the inability of our economic policies to provide an adequate income for all, to the priority placed on obtaining material goods and possessions, and to the loss of community and community control. At the root of the matter surely is that inequalities in opportunities, in education, in income, and in all other aspects of life are no longer secret and no longer accepted. Nevertheless, inequality is a fundamental and integral component of societies built on and dedicated to competition. Regardless of the nature of the race, competition means that there are winners and losers. But Canadian and American losers no longer accept this status as inevitable or necessary. Regardless of the causes, these characteristics of society set the context for social policy. It is the mission of social policy makers to respond to these conditions, to resolve and if possible to prevent them.

THE RESIDUAL APPROACH TO SOCIAL POLICY

The residual approach to social policy is distinguished by a number of characteristics. First, it largely rejects Mills' argument that there is a public issues component to social problems, despite the evidence noted above. Rather, there are a number of individuals who at a particular time experience difficulties such as an inadequate income, but who by virtue of hard work can pull themselves up by their bootstraps. The residual position vis-à-vis public issues is nicely illustrated by the story of former President Gerald Ford who, when a congressman, could be so moved by a personal story of tragedy that he was ready to literally give the shirt off his back, but who would then enter Congress and vote against Medicare. The residual vision is governed by rugged individualism and denies the case for responsible collective sharing and action.

Second, the residualist position is skeptical about the appropriateness of state intervention into social policy. It is argued that neighbors, the church, and voluntary and charitable associations are more appropriate sources of assistance, whether the need be for income, medical care, or family counselling. This position was eloquently expressed by another U.S. president, Herbert Hoover, who in the midst of the depression favored federal aid for farmers to take care of starving animals, but declared that it was up to the Red Cross and other charitable organizations to care for starving people.

Third, proponents of residualism recognize that there are some who cannot care for themselves, and that these individuals, because of sickness, age, or desertion of the family breadwinner, require aid. However, accompanying this acknowledgment is a deep-seated suspicion that there are many other individuals who do not need assistance but are eager to take advantage of programs of income assistance or medical care. Residualists fear generosity in social programs and favor the establishment of mechanisms to distinguish deserving from undeserving applicants. Despite the evidence that cheating in social welfare assistance programs is negligible,[1] residualists are driven to develop detailed sets of regulations and mechanisms for ensuring that regulations are observed. The consequence is that while benefits paid under residual programs can be meager, the costs of administration are high. Some evidence for this assertion is provided later in the chapter.

Cochran provides an accurate and insightful view of the consequence of residual social policies in the following quotation:

> The client oriented categorical approach to program eligibility and delivery is typical of human services in the United States. Experiences with the Family Matters project suggest that this is self-defeating primarily for two reasons. First, the approach attaches a stigma to the service: potential consumers immediately realize that to be associated with it they must accept an arbitrary public definition of themselves as insufficient. Those with self respect stay away from such a service and those who do enlist begin by being put down rather than uplifted. Second, such labelling takes the responsibility for identification of needs away from the consumer and places it fully in the hands of the provider shifting the consumer's role from active parties to passive recipient. This shift makes no sense if the ultimate goal is to foster independent self-supporting individuals and families. (1985, p. 23)

In summary, the residual perspective holds that state-supported services should be provided only as a last resort when the resources of individuals and families have been exhausted. Residualists such as Ronald Reagan in the United States and Margaret Thatcher in the United Kingdom are fervently committed to this perspective, believing that state-supported services rob people of their initiative and create long-term dependency and idleness. In the opinion of these politicians, the provisions of the welfare state are a problem to be eliminated, and the extent of their political support argues that they represent the prevailing opinion.

The potency of this residual approach is reinforced by the fact that these politicians share common ideologies and values with leaders in the business world who contribute substantial resources to the election campaigns of their political allies. It is obviously in

---

[1]The most recent evidence is provided by Reuben Hasson, whose national survey into welfare fraud revealed that less than 1% of all recipients of social assistance are convicted of fraud (Hasson, 1981). These findings are consistent with those of earlier investigations.

the interests of both groups to ensure that the status quo remains undisturbed. Present arrangements for earning and keeping income ensure that the distribution of income will continue to be as advantageous to the rich as it always has been. In Canada the evidence for this assertion comes from such impartial sources as Statistics Canada. In 1951 the top fifth of earners received 42.8 percent of the nation's wealth, whereas the share of the bottom fifth was 4.4 percent. In 1982 the shares of the two groups remained almost the same: 42.7 and 4.5 percent, respectively (Statistics Canada, 1982).

THE RESIDUAL APPROACH TO PRACTICE

Because of the low priority accorded to social policy, the resources allocated to income security, to health care, and to the personal social services are insufficient. The extent of the crisis-ridden reality of the state-supported mental health and social services is vividly portrayed in *Street Level Bureaucracy* (Lipsky, 1980). The reality includes clients whose lives are in such long-standing disarray that they are unable to participate as partners in the helping enterprise, by direct service staff who in many ways are as beleaguered and embedded in crises as their clients, by a chronic scarcity of resources, and by centralized, bureaucratized organizations that provide inhospitable work environments for human services workers. Given this state of affairs and recognizing that people in crisis want immediate assistance, not long-term environmental change, it is difficult for practitioners to assign time and attention to environmental issues.

A second characteristic of residualist practice flows from professional education. Few programs of professional education have developed curricula to prepare practitioners for ecological practice roles such as educator, consultant, trainer, and advocate. Some graduate programs in social and community psychology have anchored their curricula in the ecological perspective. However, in addition to being few and far between, these programs are concerned mainly with research rather than with practice. A number of undergraduate programs in social work have embraced the ecological perspective only to see it abandoned at the graduate level. Typically, graduate programs in the human services specialize, and since specializations tend to narrow rather than widen perspectives, the breadth of vision required by the ecological approach is denied by the majority of graduate programs. To compound the difficulty, graduates of masters and doctoral programs command more status and occupy the policy and administrative positions. They have considerable influence in shaping the human services in specialized rather than more broadly based ecological approaches to policy and practice.

A third characteristic is that new and difficult problems demand the emergence of specializations. Witness, for example, the absolute and very understandable sense of panic of the newly graduated child care or social worker confronting his or her first case of sexual abuse. The gravity and the complexity of the case cry out for more knowledge and additional skills—in short, for being more adept in handling a particular kind of situation. While it is true that the incidence of sexual abuse has risen dramatically in the past few years, it nevertheless remains a relatively small part of child welfare caseloads. However, because it results in long-term trauma and because of its emotion-laden quality, it tends to drive child welfare workers to demand specialized training and consultation. And in an era of scarce resources it is easier to argue for additional resources for responding to the crisis of sexually abused children than for changing conditions in neighborhoods.

A fourth characteristic is that while it is appropriate and necessary that human services professionals aid individuals in distress, it is difficult for them to engage in social change. Public issues such as substandard housing, the lack of recreation facilities, and inadequate incomes are seen as the preserve of politicians, not professionals. In addition, the task of organizing neighborhood or client groups and documenting social conditions is often seen as inappropriate for professionals employed directly or even supported by the public purse.

The above characteristics can be viewed as powerful obstacles standing in the way of an ecological approach to practice in the human services. As noted here, these include the difficulty of introducing approaches that seek to promote the competence of and empower clients, the continuing crisis-ridden nature of the service system, specialized educational programs that ill equip practitioners to practice in a holistic fashion, and the reluctance of politicians and other citizens to allow human services workers to engage in social as opposed to individual change.

THE UNIVERSAL APPROACH TO SOCIAL POLICY

This section of the chapter identifies policies and practices that are compatible with the ecological perspective. The opposing approach to residualism is the universal approach which can be traced back to the Beveridge Report in the United Kingdom (1944). In that country its most articulate and committed advocates have been Richard Titmuss (1968) and Brian Abel Smith (1968). In Canada the early work of Leonard Marsh (1975) and Harry Cassidy (1943) was extremely influential in developing universal programs, and in the United States the writings most often referred to are those of Harold Wilensky and Richard Lebeaux (1958), David Gil (1976), and Alfred Kahn and Sheila Kamerman (1983). Put most simply, the universal approach advocates that income replacement programs, health care, and the personal social services should be organized as social utilities— as expected and normal provisions for individuals and families.

> This definition implies no stigma, no emergency, no abnormality. Social Welfare becomes accepted as a proper legitimate function of modern industrial society in helping individuals achieve self fulfillment. (Wilensky & Lebeaux, 1958, p. 140)

The universal approach acknowledges that the state has a responsibility to address public issues such as unemployment, and problems caused by the lack of income, illness, and poor housing. Further, it seeks to address these problems in a way that does not denigrate and stigmatize those who apply for aid. Rather, it seeks to promote opportunities for individuals and families and in its most extended and ambitious conception, that of social development, seeks to involve individuals in the planning and management of programs that affect them.

The essential characteristics of the universal approach can be demonstrated by some examples of programs in three fields of service: income security, health care, and the personal social services. Canadian examples of universal income security programs are the demogrant programs of Old Age Security and Family Allowances. Both of these demogrant programs pay benefits to individuals in certain age categories as a right and regardless of income. Hence individuals are not required to prove that they need or deserve benefits, and since all receive benefits, no stigma or declaration of failure is attached to the program or to recipients. The rationale for family allowances is that raising children is a costly enterprise, and one in which the country as a whole has a stake and a responsibility. Thus the state seeks to assist in this expensive endeavor by making a financial contribution. In the same way, since retirement results in reduced income for all citizens, the old age pension recognizes this reality.

The major criticism of the demogrant programs is that they are wasteful and target inefficient. Why pay family allowances and old age pensions to the rich, who do not need the money, and to the welfare cheats, who do not deserve them? The following brief comparison between universal and residual approaches to income security and medical care addresses these criticisms.

The difference between universal and residual provisions for income security is strikingly evident in the policies of the United States and Sweden. The former country relies heavily on means-tested programs (33% of the cost of all income programs) and has no demogrants. On the other hand, Sweden's proportional expenditures are 42 percent on demogrants and only 10 percent on means-tested programs.

Berglin and Hokenstad conclude their review of the two approaches with the following comment:

> It is clear that the current emphasis on means-tested welfare exacerbates "we–they" tensions in American society. Coupled with trickle-down economic policies, the current approach to welfare fosters the continuation of an underclass which both alienates and is alienated from "productive" groups within the society. Reliance on means-tested benefits contributes to and reinforces the divisions of slum and suburb, poverty and wealth, weak and strong which tear at the fabric of the society. A policy designed to provide income support to most Americans based on demographic characteristics would focus attention on mutual benefits rather than on social and economic disparities. In doing so, it would help to lessen the "we–they" tensions so evident in the U.S. today. (1981, p. 85)

Studies of income security programs in Canada make the point that demogrants are cheap to administer and free of stigma, while the administrative costs of selective programs are exorbitantly high. Michael Mendelson estimated that in 1973–74, $450 million could have been saved if income replacement programs had taken the form of "a single demogrant providing fixed monthly payments to every man, woman and child" (1979, p. 86). Irwin Gillespie concluded in his review of income security programs that in the 1970s

> the federal government has not substantially improved the economic position of the poor relative to the highest income families. . . . Further this study pinpoints the few policies that are redistributive: Old Age Security, the Guaranteed Annual Income Supplement, Family Allowances. (undated, pp. 3, 25)

There is then some evidence that argues that demogrants do distribute income from the rich to low-income groups, that they are relatively cheap to administer, and that they do not stigmatize recipients. Further, it should be added that some recent innovations in income security in Canada have centered around the tax credit, whereby individual families receive benefits from the tax system if their income falls below a certain level. The use of income tax and its impersonal mechanism for distributing benefits departs from the tradition of demogrant payments, but is nevertheless relatively easy to administer and free from stigma.

In health care the universal approach assigns the responsibility for covering the costs of hospital and outpatient medical care to the state, whereas under a residual approach the individual is held responsible. While the former is often castigated for promoting dependency on the part of consumers (why not call on the family M.D. if the visit is free?) and for being wildly expensive, the evidence does not support these assertions. John Calvert's conclusions agree with many other inquiries into the comparative health care costs between the universal approach in Canada and the residual position of the United States.

> With regard to the questions of how efficiently each system uses its resources, a strong case can be made that the Canadian system provides a much better quality of service, despite the lower level of funding it receives. Overhead administration costs amount to only 2.4% of total spending on health care in Canada, compared with 12.5% in the U.S. The Canadian health care system is by no means perfect. But for the average working person the availability and quality of health care is significantly better than in the U.S. (Calvert, 1984, p. 111)[2]

The differences are equally evident with respect to the personal social services—those designed to provide counselling, home help, and day care programs to families. Here the residual perspective essentially holds that most families are sturdy and self-reliant, and can function quite capably with support from friends, neighbors, and relatives.

The universal approach is predicated on the assumption that all families experience some difficulties, particularly at times when roles and responsibilities change—at transition points such as marriage, the birth of children, when children leave home, separation and di-

---

[2]See also the work of Robert Evans (1983).

vorce, and death. Given the normalcy of these transitions it makes sense to provide services that will aid and support families at these times. The universal approach argues for the establishment of personal social services as social utilities for families. These should be readily accessible (in the immortal words of the Seebohm Commission, in "pram-pushing distance") and free from stigma.

To summarize, the universal approach calls for adequate incomes, health care, housing, and support services for families. These should be available in nonstigmatized locations and provided in such a way as to enhance dignity and promote independence. The argument being developed here is that implementation of the ecological perspective demands universal policies to address public issues and to provide resources for preventive as well as responsive programs.

A Preventive Approach to Practice

Enhancing Competence

It is difficult to provide conclusive evidence that practice approaches based on the ecological perspective are effective in both cost and human terms. But some scattered evidence is at least suggestive and is reported in detail in a later section. Two examples of projects that have enhanced competencies are noted here. A project that assigned public health nurses to undertake postnatal visits to provide training for mothers deemed to be at risk revealed a much reduced incidence of child abuse than occurred in a similar group of mothers where no postnatal visiting occurred (Gray, Cutler, Dean, & Kempe, 1979). Although early findings were not encouraging, the U.S. Head Start program has been carefully evaluated, with the following conclusion:

> Head Start can increase young children's readiness for school. It can provide some useful social and intellectual skills to support children throughout the school years, leading to better performance in the later grades. Head Start can deliver to parents important social and educational services that they can use to reinforce their children's learning experiences. (Valentine & Zigler, 1983, p. 278)

Changing Negative Environments

An example of effective environmental change was accomplished by a community association in Chicago that undertook an analysis of the reasons for residents being hospitalized and being treated at the emergency clinic. The association discovered that the following accounted for the majority of the hospital visits: auto accidents, interpersonal attacks, other accidents, bronchial ailments, alcoholism, drug-related problems, and dog bites. Through this analysis, the association recognized that "the sicknesses" affecting the community had little to do with the medical expertise of the hospital, and that the appropriate remedies lay in changing traffic patterns, improving street lighting, developing family counselling programs and A.A. groups, and catching stray dogs!

> This example of community health action suggests that improved health is basically about moving away from being "medical consumers." Health is a political question. The health action process can enable health development by translating medically defined problems and resources into politically actionable community problems. (McKnight, 1978, p. 39)

The action of this association improved the health of the community and saved money. Even the most committed residualist would have to applaud. However, engagement in community change is often seen as pushing staff from appropriate professional practice into political activities. One alderman wrote to the *Journal of Ontario Children's Aid Society* in response to an article by the writer urging the CAS become involved in community work.

> Community work tasks of need identification, mobilizing groups, assisting them to consider alternative courses of action and then developing programs are political not professional tasks. If a social worker feels strongly about what ought to be done let him enter politics directly; he has no right to expect a salary from the taxpayer while operating a political organization which he has built up and which from behind the scenes he directs and controls. (MacDougall, 1972, p. 17)

The issue is an important one and one that ecological theorists and community development leaders have been negligent in addressing. For this writer the issue cannot be resolved by one deft delineation of boundaries between professional and political jurisdictions. Rather the boundary is permeable and will shift depending on the actors and the community in which the action is being carried out. But a beginning can be made by recognizing and adhering to the mandate of social service organizations. If the mandate is to prevent the neglect and abuse of children, or to prevent mental illness, and if professional staff can demonstrate that developing the competence of families, building local-level resources, and changing neighborhood conditions can contribute to the mandate, the activity can be supported. This includes the documentation of conditions such as poverty and poor housing, and there are examples of Children's Aid Societies in Ontario that have made a commitment to document adverse conditions and to advocate for change on behalf of clients.

Empowering People to Take Charge

> Human services have come full circle reaching back to the beginnings of organized human services to find the basis for a renewed partnership with people who are helping one another in their everyday lives. This shared philosophy emphasizes the principles of self determination, self reliance and mutual aid which serve as a frame of reference for staff in providing help. (Froland et al., 1981, p. 55)

The essence of this partnership model can be found in examples of community work approaches to child welfare in Native Indian communities where professional staff have focussed their efforts on developing resources in the reserve, enhancing the competence of families to parent, and empowering the members of the band or tribal council to own the problems of child neglect and abuse (Wharf, 1985a).

Thus, the concept of empowerment can be extended to the level of communities and can be used as an argument for community control over human services. The teachings of political theorists such as Carole Pateman suggest that the case for participatory democracy begins in local communities and in the workplace, on the rationale that these spaces provide a useful training ground for citizens to gain competence in the political process (Pateman, 1970). From his review of the Family Matters Project, Cochran argues that

> the empowerment process consists of a series of stages involving transactions by the individual with progressively more distant environmental systems. Positive changes in self perception (Stage I) permit the alteration of relations with members of the household or immediate family (Stage II), which is followed by the establishment and maintenance of new relations with more distant relatives and friends (Stage III). Stage IV is seen as information-gathering related to broader community involvement followed in Stage V by change-directed community action. (1985, p. 25)

The process of empowerment can lead to the control of, or at least some significant share in, the ownership of community-based and relevant services such as health, education, and social services. While some forms of community ownership of health and education have occurred in Canada, this has not occurred in the social services. The case for community control is argued forcefully in a recent review of the social services in British Columbia. *Reforming Human Services* advocates a participatory model between province and communities. The division of responsibility envisioned in the book sees the provincial government as responsible for establishing legislation, setting budgets, developing standards, and operating some specialized province-wide services (Clague, Dill, Seebaran, & Wharf, 1985).

In this view, communities should be responsible for planning and delivering services to families and children in accordance with a plan that would specify the kinds of services required in the community and the most suitable structures for governing and managing the services.

To conclude this discussion of the practice component of the ecological perspective, it has been argued that professionals can enhance the competence of clients and can assist them to have more control over their lives. Further the potential exists for community control of the personal social services, which in turn paves the way for professionals and consumers to contribute to the development of social policies.

TOWARD IMPLEMENTATION OF THE ECOLOGICAL PERSPECTIVE

Implementing the ecological perspective will be exceedingly difficult. The most fundamental obstacle is of course that policy makers in the United States and Canada are committed to a residual approach both in practice and in policy. Given their commitment to this position, they are unwilling to assign additional resources to the social services and by so doing alleviate some of the crisis-ridden reality of practice. In addition, their suspicion of social services and their conviction that clients take advantage of the services is such that they seek to regulate and control services, thus creating the bureaucratized and inhospitable work environment described earlier. At the present time Canada and the United States are characterized by politics of exploitation, which aid industry and the rich generously and with few restrictions, but which are meanspirited and restrictive in providing assistance to the poor. The syndicated columnist Ellen Goodman (1981) puts the point succinctly.

> According to the politics of the new fiscal year, the rich have lost their willingness to work hard because the government has taken too much money away from them. The poor, on the other hand, have lost their willingness to work hard because the government has given too much money to them. In response to this grave situation as of October 1st we have cut the taxes of the rich most lavishly, giving them more money and more incentive to work. We have cut aid to the poor, giving them less money and therefore more incentive to work . . . now no one has explained to me exactly why the rich need money to make them labour while the poor need desperation. (Goodman, 1981, p. 5)

Despite these formidable obstacles, a number of strategies for change can be suggested. First, the crisis-ridden reality of practice continues because the potential of the ecological approach to save money has not been documented with sufficient care. In a review aimed at determining the differences between funded and approved prevention programs and rejected proposals in the United States, Pauline Pizzo argues that the cost saving elements of a proposal have become virtually central to its successful enactment.

> Between 1976 and 1980 four programs for young children were, in comparative terms, substantially and deliberately expanded. WIC (high-protein supplemental foods for mothers and babies), adoption assistance and services to prevent long-term foster care placement, childhood immunization, and Head Start. Supporters of all four programs were able to present a persuasive case that without this expansion more costly expenditures would have resulted. Other proposals—no matter how meritorious in other respects—failed to win the same support. (Pizzo, 1983, p. 29)

EXAMPLES OF COST–EFFECTIVE PROGRAMS

In an attempt to locate examples of cost-effective programs, the writer undertook a follow-up of 96 projects reported in the Annotated Directory of Family Based Service Programs compiled by the National Resource Centre of Family Based Services. The following chart indicates the breakdown of the responses:

Number of responses to the initial and
a second letter   44
Number where evaluation had not been
completed   14
Number of evaluation reports considered
to be lacking in basic data   12
Number of projects with a complete and
adequate evaluation component   18

The common objective of all projects was to prevent the out-of-home placement of children considered to be at risk by professional child welfare and other human service workers. While few projects utilized an experimental design, all carefully tracked the outcomes of providing in-home services (counselling, therapy, homemakers, parent education) for families characterized by inadequate income and a history of parent–child difficulties. With only one exception, the outcomes ranged from a success rate of 68 percent to 98 percent. The differences were attributable to the number of programs available and the adequacy of the resources supporting these programs.

Only 5 of the 18 successful projects reported on the financial implications of home-based services. These are:

1. *The Placement Alternative Program*, Dakota County, Minnesota, reduced a $323,174 foster care budget in 1980 to $273,774 in 1982.
2. *The Intensive Probation Supervisor Project*, Ramsay County, Minnesota, reported an annual $1 million saving by assisting two-thirds of the youths released from a correctional center to remain at home.
3. *The Homemakers Program* in Seattle indicated that their services cost $1000 less per child per year as opposed to foster care.
4. *The Boulder Department of Social Services* in Colorado estimated that their programs resulted in a $430 reduction in costs per month per child.
5. *Conserving Families* in Burlington, Iowa, cut the costs of foster care by 23% and of group home care by 39%.

The above results are consistent with those obtained in more rigorous studies (Seitz et al., 1985; Olds, in press; Gray et al., 1977) concerned with the prevention of abuse and neglect. The available evidence is that programs that support families can reduce out-of-home placements and that such programs are cost-effective.

A second strategy to implement the ecological perspective involves changes in the education of human services workers and in the roles and responsibilities they are assigned in practice. The change involves preparing professionals in the human services to assume responsibility for functioning as teachers, consultants, and coaches. For too long, professional education and practice have stressed the role of the expert as a clinician, managing and manipulating the behavior of clients. These roles need to be balanced by providing information and teaching skills, and promoting the formation of self-help and support groups—strategies that enhance the competence of clients and lead to empowerment.

This shift has occurred in some agencies and projects, and the consequences for professional practice and service to clients are nicely captured in a report of the Parent to Parent project of the High Scope Educational Research Foundation of Ypsilanti, Michigan.

> It may appear that we no longer value professional contributions and expertise. This is not the case. Instead, the professional's role changes. Professionals become more effective in training and supervisory roles and are thus able to use their knowledge to benefit even more people than they can when they work in one-to-one relationships. Further, professionals are freed to use their expertise helping severely dysfunctional families who require skilled assistance beyond that of our trained professionals.
>
> As professional roles change, shifts in attitude also occur, gradually transforming the traditional hierarchy of service-provider roles: Families become active participants in change rather than dependent recipients. Volunteers and paraprofessionals are viewed as skilled individuals, providing services in exchange for training and institutional support, rather than "cheap labor." As supervisors and trainers, professionals use their expertise and knowledge to develop resources and support for families working to help themselves. They are no longer direct service providers trying to bridge the gap between their own values, backgrounds, and training, and the lives of families they served. Educators, researchers, and program directors become partners with the community by translating child development information and experience into a program that develops community child-rearing competence. (Evans et al., 1984, p. 10)

There really is no disagreement between residualists and professionals of an ecological persuasion that when natural helping from kin and friends can occur, it should; where it can be nourished by professionals, this is an appropriate professional task; and where participation between professionals and natural helping networks can be developed, this should be encouraged. It is agreed that individuals should be resourceful and as self-sufficient as possible and should contribute to the community.

However, ecological theorists and practitioners have not altered the image of human services professionals that persists in the minds of residualists—namely, that professionals are interested primarily in their own welfare and seek to promote this by developing and maintaining their positions as experts through the use of esoteric skills and jargon. This is not how ecological practitioners view their mission and themselves, but communicating this to policy makers will require attention to language and to building connections where few have existed in the past.

A third strategy is to recognize and capitalize on pressure group tactics by groups such as senior citizens. In Canada the actions of these groups effectively blocked the plans of the federal government to de-index Old Age Pensions. With expanding membership and a corresponding increase in their political clout, the Gray Panther, Women, and Indian movements may become a powerful force for political change in Canada and other Western democracies. Needed is a way of coordinating the social policy agendas of these movements in order to achieve the maximum impact.

Fourth, impetus may come from the changes in the birthrate in Western society.

> Between 1968 and 1985, fertility rates of the industrial democracies tumbled far below the Zero-Population-Growth (ZPG), or replacement level. In several lands, actual population decline set in, with deaths exceeding births. In demographers' terms, a total fertility rate of 2.1 babies per woman insures the population will stay level over the long run, discounting immigration. As of 1983, West Germany and Denmark each had a rate of 1.3, the Netherlands and Italy 1.5, Japan 1.7, France and the United Kingdom 1.8. In Canada the rate fell from 2.8 in 1966 to 1.7 in 1981. In the United States, the rate tumbled from 3.6 in 1955 and 2.9 in 1965 to 1.7 in 1976, and has since hovered around that figure. Even this number significantly relies on the astonishing fertility of unmarried teenagers, a fact which comforts no one. (Carlson, 1986, p. 5)

The falling birthrate has already prompted concern. In France President Mitterand has argued that "the decline in the birthrate constitutes a grave menace for the west and we must take action" (Carlson, 1986).

Those with a long social policy memory will recall that this was precisely one of the reasons that several countries, including Canada, introduced the universal family allowance payment. As concern for the falling birthrate coupled with concern for other social problems such as child abuse, the growth of one-parent families living in poverty, and the incidence of pregnancies of teen-age mothers begins to register on the social policy agenda, federal and provincial cabinets may consider establishing universal social policies for families. It may well take the combined impact of all these forces to bring politicians, most of whom are middle-aged and wealthy men, to recognize that, like other neglected natural resources, Canadian and American children are also becoming an endangered species.

Fifth, ecological theorists and practitioners must develop social policy agendas. Some substantial work has already been published by scholars such as Sheila Kamerman, Alfred Kahn, Urie Bronfenbrenner, Edward Zigler, David Gil, and many others. Essentially their work calls for a social development approach to social policy. A social development approach requires at a minimum:

> opportunity for adult males and females to earn adequate incomes in employment that is rewarding;
> adequate housing;
> adequate medical care coverage for all families;
> adequate income support and supplementation programs for all families;
> adequate programs of day care and support services for children and families.

The above recitation is familiar. What is unique about the social development approach is an insistence that economic and social policy must be considered equal, rather than the former as predominant and the latter as picking up the unforeseen consequences. The President of the Institute for Research in Public Policy in Canada argues that Canada needs a new social contract that would require

> an integrated approach to economic and social policy recognizing that the two are mutually conforming not competing; guaranteed annual income and the need for social investment in human capital formation—health, education, social services, training, adjustment assistance, day care, conservation co-ops and the like and

in renewable resources—fisheries conservation, silviculture, social conservation, water quality—as a legitimate charge on the
public purse. (Dobell, 1987, p. 11)

An equally radical component of the social development approach is the insistence that citizens must be allowed to participate
in structures that affect them. Social development therefore pushes
for a participatory approach to democracy, and one useful beginning is to involve consumers, citizens, and professionals in the policy-making process for the social services.

But while serious and scholarly calls for reform have been developing in the literature and in the professions receptive to ecological thought and practice, these proposals have not registered with
policy makers. Edward Zigler notes his concern about this in the
following fashion:

> On the basis of my 25 years of effort, I can tell you that our many
> failures can be traced to a single fact: we have never in this coun-
> try been able to put together a broad-based, truly effective advo-
> cacy or lobbying group whose central goals are a better life for
> children and families. If all the family support programs in this
> nation shared a common vision, they could eventually be unified
> into that potent political force that has so long been missing. The
> future will tell us whether we are successful in constructing such
> a force on the American social scene. (1986, p. 12)

As noted throughout the chapter, the task of developing a common vision is fraught with difficulties. Certainly we do not expect
advocates of Pro-life and Pro-choice to agree and neither do we expect Gloria Steinem and Phyllis Schafly to see eye to eye. But by using a bottom-up approach that identifies the characteristics of
effective practice and then constructs policies from practice, the beginnings of a common vision can be developed. For family and child
services:

1. We can begin by demonstrating that programs of support to families are effective. They are not primary prevention programs and
   do not prevent difficulties and breakdowns in families from arising. But, in Cowen's words, they are effective baby steps in that
   they reduce the stress in and promote competence of high-risk
   families and thereby reduce the necessity for out-of-home care.
2. From this growing body of evidence, we can argue that state/
   provincial policies are required to put family support programs

in place, and not just on a project-by-project basis. This will allow experience to be cumulative across state and provincial jurisdictions.

3. We can agree, too, that in large part the need for family support programs is to offset the negative effects of poor environments caused by unemployment and inadequate income, housing, and medical care. It follows that if these causative factors are reduced, such programs could become more explicitly concerned with building strengths and competencies rather than offsetting negative conditions.

> Studies of neighbourhoods and families both indicate that while economic deprivation does indicate the risk of child maltreatment, it is not determinant and can be overcome. And yet we should not too readily discard the hypothesis that sustained widespread prevention will only come as a feature of efforts such as reducing poverty, improving health care, and improving the political climate for children's issues. (Garbarino, 1986, p. 155)

4. We can identify the characteristics of local-level programs that make for effectiveness. These include competence promoting and empowering methods of helping, the absence of stigma and of procedures and auspices that designate the recipients as failures and rejects. There is a compelling need for practitioners to contribute their unique experience and insights to the development of policy. A number of recent articles and research studies have argued convincingly that, particularly in the implementation stage of the policy process, practitioners play a crucial role in determining whether policies are implemented as intended (Lipsky, 1980; Wharf & Callahan, 1984; Williams, 1980).

5. Extrapolating from these characteristics, it can be argued that policies with the same characteristics would also be effective, and indeed from the Canadian experience this does appear to be the case.

Richard Elmore has described this approach to policy development as backward mapping (1982). It begins with practice, identifies what works in practice, and builds the case from the bottom. It has the advantage in this arena of starting with examples of effective practice with which even residual policy makers can agree. It then asks them to follow the logic of the approach through community and state levels to that of national policy. In this fashion, backward mapping would argue vis-à-vis Gerald Ford that health care practice that focussed on education and on health promotion activities was consistent with and most likely to flourish under a nationwide system of universal health care.

CONCLUSION

To conclude, the ecological perspective has much to commend it in both policy and practice. As emphasized throughout this chapter, one of the perspective's distinctive characteristics is its insistence on the connections among individuals, families, communities, and national social policies. While the connections have been identified in research studies, we have not yet found a way to monitor and report connections as an integral component of the human services. By and large, human services agencies are so consumed by crises and so controlled by the governments that fund them that they do not track the impact of economic and social policies on communities and families, nor are they able to bring to the attention of policy makers the results of successful empowering projects at the community level.

Zigler is correct that we need a national constituency for families. We also need provincial and community constituencies in order to develop provincial legislation that will require that attention be given to prevention and that human services agencies monitor and report on the impact of economic forces and social policies. Whether this monitoring function should be assigned to direct service agencies is a moot question. Perhaps community planning agencies are most appropriate to undertake this complicated task, and certainly the ability to report on public issues requires protection from being absorbed in crises and being directly controlled by government. The development of children's defense funds in many states in the United States is potentially promising, providing, of course, that these funds have sufficient autonomy. In particular, ecological practitioners must begin to present arguments on the grounds of cost-effectiveness. If universal health care is cheaper, if family support programs reduce the number of children having to be placed in substitute care and hence save money, these are powerful arguments in the days of understandable concern about national debts.

Finally, although not all ecological theorists and practitioners would, this writer agrees with David Gil's statement that "the primary requirements of primary prevention of child abuse amount to fundamental philosophical and structural changes of the prevailing social, economic and political order (Gil, 1976b, p. 34). It requires establishing a caring, cooperative society, which in turn requires fundamental changes such as narrowing the income gap between the rich and poor and extending opportunities for the poor to compete on a more equal footing in our essentially unequal society. It requires redistributing not only wealth but power. It requires examination of such issues as public versus private ownership of land, a guaranteed minimum income, limits on maximum incomes, and tax policies and tax write-offs. Such issues rarely get on the public agenda simply because they threaten to disturb the status quo that benefits those in power and their friends. In the last analysis, such changes in political and economic power seem possible only in small, homogeneous societies, and even the most optimistic would acknowledge that they constitute a long-range agenda for change in Canada and the United States.

REFERENCES

Abel Smith, B. (1968). *In search of justice.* London: Allen and Unwin.

Barr, D. (1979). The Regent Park Community Services Unit: Partnership can work. In B. Wharf (Ed.), *Community work in Canada*. Toronto: McLelland & Stewart.

Berglin, H. & Hokenstad, M. Jr. (1981). Sweden's demogrants: A model for the U.S.? *The Journal of The Institute for Socio Economic Studies, VI*(3), 85.

Beveridge, W. H. (1944). *Full employment in a free society.* London: Allen & Unwin.

Bronfrenbrenner, U. (1979). *The ecology of human development: Experiments by nature and design.* Cambridge, MA: Harvard University Press.

Bronfrenbrenner, U. & Weiss, H. B. (1983). Beyond policies without people: An ecological perspective on child and family policy. In E. Zigler, S. L. Kagan, & E. Klugman (Eds.), *Children, families and government* (p. 405). Cambridge: Cambridge University Press.

Calvert, J. (1984). *Government limited.* Ottawa: The Canadian Centre for Policy Alternatives.

Carlson, A. (1986). *Stakes high as birthrates collapse. Washington Post.* Reprinted in *Victoria Times/Colonist.*

Cassidy, H. (1943). *Social security and reconstruction in Canada.* Toronto: Ryerson Press.

Clague, M.; Dill, R.; Seebaran, R.; & Wharf, B. (1985). *Reforming human services: The experience of the Community Resource Boards in B.C.* Vancouver: University of British Columbia Press.

Cochran, M. (1985). The parental empowerment process: Building on family strengths. In J. Harris (Ed.), *Child psychology in action: Linking research and practice.* London: Croom Helm.

Cowen, E. (1977). Baby steps toward prevention. *American Journal of Community Psychology, 5,* 1–22.

Dobell, R. (1987). The continuing challenge of full employment. Notes for an address to the Canadian Clubs in Alberta, Institute for Research in Public Policy.

Elmore, R. (1982). Backward mapping: Implementation research and policy decisions. In W. Williams (Ed.), *Studying implementation.* Chatham, NJ: Chatham House.

Evans, J., Walker, S., Parker-Crawford, F., de Pietro, L. and Epstein, A. S. (1984). *Roots and wings: Parent to parent dissemination* (p. 10). High Scope Educational Research Foundation, Ypsilanti, MI.

Evans, R. (1983). Health care in Canada: Patterns of funding and regulation. *Journal of Health Politics, Policy and Law. 3*(1), 1–43.

Froland, C.; Pancoast, D. L.; Chapman, N. J.; & Kimboko, P. J. (1981). *Helping networks and human services.* Beverly Hills, CA: Sage Publications.

Garbarino, J. (1986). Can we measure success in preventing child abuse? *Child Abuse and Neglect. The International Journal, 10*(2), 143–155.

Garbarino, J., Stocking, S. H., Collins, A. N., Gottlieb, S., Olds, D. L., Pancoast, D. L., Sherman, D., Tietjen, A. M., & Warren, D. I. (1980). *Protecting children from abuse and neglect.* San Francisco: Jossey-Bass.

Germain, C. & Gitterman, A. (1980). *The life model of social work practice.* New York: Columbia University Press.

Gil, D. (1976a). *The challenge of social equality.* Cambrige, MA.: Shenkman.

Gil, D. (1976b). Primary prevention of child abuse: A philosophical and political issue. *Psychiatric Opinions, 13*(2), 30–34.

Gillespie, I. (undated). *In search of Robin Hood: The effect of federal budgetary policies during the 1970s on the distribution of income in Canada.* Montreal: The C. D. Howe Research Institute.

Goodman, E. (May 5, 1981). The two track work ethic. *Victoria Times.*

Gray, J. D.; Cutler, C. A.; Dean, J. G.; & Kempe, C. H.; (1979). Prediction & prevention of child abuse and neglect, *Journal of Social Issues, 35*(2).

Guberman, N. (1987). State–family relations: Privatizing social services. *Perceptions, 10*(3), 16–18.

Hasson, R. (1981). The cruel war: Social security abuse in Canada. *Canadian Taxation: A Journal of Tax Policy, 3*(3), 114–147.

Kahn, A. J. & Kamerman, S. B. (1975). *Not for the poor alone.* Philadelphia: Temple University Press.

Kamerman, S. B. & Kahn, A. J. (1983). Child welfare and the welfare of families with children: A child and family policy agenda. In B. McGowen & W. Meezan (Eds.), *Child welfare: Current dilemmas and future directions.* Itasca, IL: Peacock Publishers.

Lipsky, M. (1980). *Street level bureaucracy.* New York: Russell Sage Foundation.

MacDougall, L. (1972). Letter to the Editor. *Journal.* Ontario Association of Children's Aid Societies, 9(15), 17.

McKnight, J. (1978). Politicizing health care. *Social Policy*. November/December, 36–39.

Marsh, L. C. (1975). *Report on social security for Canada*, 2nd ed. Toronto: University of Toronto Press.

Mendelson, M. (1979). *The administrative costs of income security programs: Ontario and Canada*. Toronto: Ontario Economic Council.

Mills, C. W. (1959). *The sociological imagination*. New York: Grove Press.

National Council of Welfare. (1975). *Poor Kids*. Ottawa: NCW.

National Council of Welfare. (1978). *The hidden welfare system*. Ottawa: NCW.

National Council of Welfare. (1979). *The hidden welfare system, revisited*. Ottawa: NCW.

National Council of Welfare. (1983). *Family allowances for all*. Ottawa: NCW.

Olds, D.; Chamberlin, R.; Henderson, C.; & Tetelbaum, R. (in press). The prevention of child abuse and neglect: A randomized trial of nurse home visitation. *Pediatrics*.

Pancoast, D. L. (1980). Finding and enlisting neighbours to support families. In J. Garbarino, H. Stocking, & Associates (Eds.), *Protecting children from child neglect and abuse*. San Francisco: Jossey-Bass.

Pateman, C. (1970). *Participation and democratic theory*. Cambridge: Cambridge University Press.

Pelton, L. (Ed). (1981). *The social context of child abuse and neglect*. New York: Human Science Press.

Pincus, A. & Minahan, A. (1973). *Social work practice*. Itasca, IL: Peacock Publishers.

Pizzo, P. (1983). Slouching toward Bethlehem: American federal policy perspectives on children and their families. In E. F. Zigler, S. L. Kagan, & E. Klugman (Eds.), *Children, families and government* (p. 29). Cambridge: Cambridge University Press.

Statistics Canada. (1982). *Income distribution by size* (Catalog 13–207). Ottawa: Statistics Canada.

Schwartz, W. (1969). Private troubles and public issues: One social work job or two. In *The social welfare forum*. New York: Columbia University Press.

The Seebohm Report. (1968). *Report of the committee on local authority and allied personnel services*. London: Her Majesty's Stationery Office.

Seitz, V.; Rosenbaum, L. K.; & Apfel, N. (1985). Effects of family support intervention: A ten year follow-up. *Child development, 56*(2), 1.

Titmuss, R. (1968). *Commitment to welfare*. New York: Pantheon Books.

Valentine, J. & Zigler, E. F. (1983). Head Start: A case study in the development of social policy for children and families. In E. F. Zigler, S. L. Kagan, & E. Klugman (Eds.), *Children, families and government* (p. 278). Cambridge: Cambridge University Press.

Warren, D. (1980). Support systems and different types of neighbourhoods. In J. Garbarino, S. H. Stocking, & Associates (Eds.), *Protecting children from abuse and neglect*. San Francisco: Jossey-Bass.

Wharf, B. (1985a). Toward a leadership role in human services: The case for rural communities. *The Social Worker, 53,* 14–20.

Wharf, B. (1985b). Preventive approaches to child welfare. In K. Levitt & B. Wharf (Eds.), *The challenge of child welfare.* Vancouver: University of British Columbia.

Wharf, B. & Callahan, M. (1984). Connecting policy and practice. *Canadian Social Work Review,* pp. 30–52.

Whittaker, J. (1983). Mutual helping in human service practice. In J. Whittaker & J. Garbarino (Eds.), *Social support networks.* New York: Aldine.

Wilensky, H. & Lebeaux, R. (1958). *Industrial society and social welfare.* New York: Free Press.

Williams, W. (1980). *The implementation perspective.* Berkeley: The University of California Press.

Zigler, E. F. (1986). The family resource movement: No longer the country's best kept secret. In *Family resource coalition report, 5*(3).

CONCLUSION

ALAN R. PENCE

Robert Glossop in the opening chapter describes the effect of Bronfenbrenner's work in the 1970s as the equivalent of a call to let a "hundred [intellectual] flowers bloom." As these chapters testify, the field of child and family research was fertile ground for such a summons. After decades of developmental research dominated by reductionistic principles, psychology and other disciplines concerned with human development were ready to explore an approach to research that was more holistic in its conception. The direction indicated by Bronfenbrenner was not a new one, but rather a "road less travelled by," and for a new generation of researchers it had the fresh appeal of a forgotten path.

These 10 chapters are a sample of the blossoms that emerged from Bronfenbrenner's and others' call. They are evidence both of the readiness of the field and the diversity of pioneers who pointed out the ecological path. Yet despite the geographic, the historic, and the disciplinary diversity of the perspectives presented, there are a number of unifying themes that thread through and make of them an ecological strand. Elements within the various chapters will be considered as they contribute to an ecological approach.

Central to an ecological orientation is the concept of interrelationships between systems extending from the most encompassing of societal influences to the most individual of behaviors. This theme emerges in most of the chapters, but perhaps Wharf's quotation from the Seebohm Commission brings the connection between macrolevel social policy decisions and micro-level individual behavior most vividly to mind. Wharf notes the Commission's policy decision that services must be delivered "within pram-pushing distance" of the individual.

Bronfenbrenner's own effective image of the relationship between system levels is contained in his nested-dolls image of the micro, meso, exo, and macro levels, each imbedded within the other.

Again, elements of this specific ecological model are evident within most of the chapters, but Shera's study of the planned development of Tumbler Ridge represents one of those rare opportunities where all of the system levels can be considered as part of the community creation process. The ability of the ecological model to complement and extend other community planning models is apparent in Shera's chapter.

Goelman and Pence's work with the Victoria Day Care Research Project demonstrates another application of the nested-systems model, this time with a focus on the mesosystem interaction of home and day care site. Unlike some areas of child and family study, day care has received considerable research attention since the early 1970s; however, the majority of those studies have tended to follow a microsystem approach in their research design, focussing, for example, on type of care as the independent variable and child development measures as the dependent variable. Utilizing a mesosystem design, the Victoria Project was able to discern critical interactional elements between the two microsystems of family and day care that help to explain earlier discrepant findings in the literature and contribute to a more holistic understanding of the complex world of day care.

Johnson and Abramovitch similarly moved beyond a microsystem study of paternal unemployment to consider the mesosystem of employment and family life. Information gathered from such interactions between systems is generally more instructive for intervention and social policy planning than is the more restricted information gathered from monosystem designs.

While the Victoria Project and Johnson and Abramovitch employed "naturally" occurring subject groups (employed/unemployed, type of care users, etc.) to order their research design, both Cochran's and Powell's projects demonstrate the power of ecological orientations in formulating intervention studies. Cochran's Family Matters Project, conceived by both Cochran and Bronfenbrenner in the mid-1970s, is not only one of the earliest but also one of the largest and most complex ecologically oriented interventions to be undertaken to date. With Cochran's work, as in the Victoria Project, the differentiability of outcomes produced by family characteristics in interaction with the intervention (either created or naturally occurring) is one of the major findings. In other words, the "same" intervention has very different impacts on different subject groups.

Insensitivity on the part of social interventionists and policy planners to such critical differences across social groups led to the "monolithic treatment" models criticized by Powell in his planning of the Child and Family Neighborhood Program. Powell's project, unlike the monolithic model, incorporated systems that would allow for the modification of the intervention if conditions or expectations within the target population called for such revisions. Indeed, the characterization of the neighborhood provided by community observers prior to the intervention proved to be different than the social structures identified by the project itself. In both Powell's and Cochran's projects, mid-stream modifications were made in order to better meet the needs of the "clients."

Sensitivity to subject perception is a key issue in the *Ecology of Human Development* (Bronfenbrenner, 1979) and one that became critical in Anglin's development of the Parent Networks Project. Anglin quotes from Bronfenbrenner: "This means that it becomes not only desirable but essential to take into account in every scientific inquiry about human behavior and development how the research situation was perceived and interpreted by the subjects of the study" (1979, p. 30). Placing himself in the role of respondent to his own questionnaire, Anglin began to question the validity of responses gained in an environment that might not be psychologically comfortable for the subject or ecologically appropriate.

The sensitivity to subjects' perceptions of research projects exhibited by Anglin, Powell, and Cochran in their interventions, and their commitment to treating the individuals involved not just as abstract "subjects"—anonymous contributors to the overall "n"—is not only an additional common theme in ecological research, but it is also a powerful bridge connecting ecological work to an enhanced understanding of the effects of social policy on individuals. As Wharf would argue, positive social policy formulation must be sensitive to the effect of macro- and exo-level decisions on the micro-level daily lives of individuals. Research that is insensitive to the personal perceptions of its subjects and that places the sanctity of research methodology above that of human experience cannot reform policy in a socially sensitive way. It is the attempt on the part of certain traditional research approaches to isolate scientific inquiry from the buzzing confusion of everyday life that both undermines the generalizability of such rigidly controlled research and draws into question its suitability as a vehicle for social policy formulation and social development. Indeed, as noted by Glossop, it is not only in the social sciences but in the natural sciences as well that a reappreciation of *context* can be found: "What is relevant *is* the pattern" (Bohm, 1981, pp. 133–34).

Anglin's concern with how one can gather data in an ecologically valid manner forms one part of Lero's discussion of national surveys. Such large scale undertakings tend to be seen as ecologically unwieldy, incapable of generating the sort of personally sensitive and qualitatively valuable information that constitutes another major dimension of ecological research. In addition to considering how such research can include ecological constructs in its design, Lero also makes the case that national survey data can provide an important contextual backdrop for smaller-scale studies. These smaller studies can in turn use a broader variety of instruments and design elements to more fully define specific parts of the larger screen.

Lamb's chapter provides an example of how such national data can provide a useful background screen for more localized studies. Having provided the macro-social context of adolescent pregnancy and parenthood (in the United States), the specific results of a Salt Lake City study provide color and detail at a more individual level. Lamb, in his juxtaposition of information from different system levels, makes one mindful of one of the pioneers of ecological thought in psychology, Kurt Lewin. Lewin's identification of subject and field interaction is a direct antecedent of Bronfenbrenner's constructs.

These chapters represent a rich diversity of ecological thought and understanding. Their roots emanate from a variety of traditions and disciplines within the social sciences. Glossop in the opening chapter traces the philosophical lineage of ecological thought in Western science; and although he mounts a tentative (and useful) "necessary . . . sufficient" set of criteria for ecological research, he stops short of a final definition, astutely avoiding the very type of reductionistic and classificatory impulses that for so long impeded the holistic, ecological pathway of inquiry. At this point in the evolution of ecological inquiry, the imposition of set and rigid criteria to "qualify" as ecological would be to stifle the call for a hundred blossoms of inquiry and restrict the field to a plot sufficient for only a few.

The chapters gathered here are not rigidly uniform in design, concept, or methodology. Indeed, some do not report on research *per se*, but instead consider the ecological framework from practice, policy, and historical-philosophical perspectives. Given such diversity of backgrounds and interests, one would not expect to find unifying threads, or a common language, that would join these chapters or that could provide an arena for meaningful interaction among the participants in their earlier workshop discussions. Yet those threads do lace through, the unifying themes emerge, and a "substantive eclecticism" is formed. Each of the chapters speaks to interactions, to reciprocal systemic influences, to the dignity and uniqueness of individuals, to the sensitive meeting of scientist and subject, and to the ultimate irreducibility of human experience.

For some the field of ecological inquiry is too broad, too encompassing, too unrestricted. But then it must be remembered that such descriptions also define the phenomena we seek to understand. For those who, above all else, require answers to their questions, who seek the bedrock of certainty in a world of simple cause and effect, the ecological route is a path of frustration, for it will certainly yield far more questions than answers. To enjoy the ecological pursuit, one must delight in the *search* of research, or as T. S. Eliot put it in *Little Gidding*:

> We shall not cease from exploration
> And the end of all our exploring
> Will be to arrive where we started
> And know the place for the first time.

For those who can delight in the process of successively understanding phenomena in new and ever different ways, who can see "a World in a Grain of Sand" and place themselves within the parameters of their own inquiry as part of the interactive dance, the ecological path is suitable, satisfying, and scientifically rewarding.

About the Editor and the Contributors

Rona Abramovitch, Ph.D., is a developmental psychologist teaching at Erindale College, University of Toronto, in Toronto, Canada. She received her doctorate from the Institute of Child Development, University of Minnesota. Her research focuses on social development and family interaction.

James P. Anglin, M.S.W., is an associate professor in the School of Child Care, University of Victoria, in Victoria, Canada. His professional interests include the future of the child care profession, family support, and parent education. He is currently studying collaborative approaches to research in the social sciences.

Urie Bronfenbrenner, Ph.D., is the Jacob Gould Schurman Professor of Human Development and Family Studies, and of Psychology, at Cornell University. Prof. Bronfenbrenner is the author of *Two Worlds of Childhood: U.S. and U.S.S.R.*, and *The Ecology of Human Development*. His research is in the areas of theory and research design through the life course, and their implications for policy and practice. He also thinks teaching is important, and likes to sing.

Moncrieff Cochran, Ph.D., is an associate professor in the Department of Human Development and Family Studies at Cornell University, Ithaca, New York, where he divides his time between research and extension work. His present interests include: the empowerment process, family support systems, informal social networks, and cross-cultural policy studies.

Robert G. Glossop, Ph.D., is a sociologist and policy analyst who serves as the Coordinator of Programs and Research for the Vanier Institute of the Family in Ottawa, Canada. His areas of research interest and publication encompass economics and family life, family policy, the integration of conceptual frameworks for family research, philosophy of the social sciences, and the design and delivery of family support services.

HILLEL GOELMAN, Ph.D., is Associate Professor of Early Childhood and Language Education in the Faculty of Education at the University of British Columbia, in Vancouver, Canada. His major research interests are in the areas of child language development, emergent literacy, and the effects of different preschool and day care settings on these aspects of child development.

LAURA C. JOHNSON, Ph.D., is a sociologist and program director at the Social Planning Council of Metropolitan Toronto. She received her doctorate from Cornell University. Her work focuses on social policy issues related to work and family life.

MICHAEL E. LAMB, Ph.D., is currently Chief of the Section on Social and Emotional Development at the National Institute of Child Health and Human Development in Washington, D.C. He has published extensively and broadly in the areas of child development and families. Lamb was formerly Professor of psychology, psychiatry, and pediatrics at the University of Utah, where much of the research described in his chapter was conducted.

DONNA S. LERO, Ph.D., received her doctorate from Purdue University. She is an associate professor in the Department of Family Studies at the University of Guelph in Canada, where she teaches courses in research methods, child development, and child and family welfare. Her major research interests are child care, child abuse, and family socialization processes.

ALAN R. PENCE, Ph.D. (Editor), is Associate Professor and Director of the School of Child Care at the University of Victoria, in Victoria, Canada. Prior to joining the University of Victoria, he was active in the provision of child care services in both the United States and Canada. His present research interests include ecological, historical, and policy perspectives on day care and working families.

DOUGLAS R. POWELL, Ph.D., is an associate professor, Department of Child Development and Family Studies, Purdue University, West Lafayette, Indiana. His current research interests focus on day care issues, parent education programs, and the professional development of child care providers. At present, Professor Powell serves as research editor of *Young Children*.

WES SHERA, Ph.D., is an associate professor in the School of Social Work at the University of Victoria in Victoria, Canada. His areas of expertise and experience include evaluation research, community development, social planning, and social impact assessment. He is currently on the council of the Canadian Evaluation Society and is an active member of the American Evaluation Association.

BRIAN WHARF, Ph.D., is Dean of the Faculty of Human and Social Development at the University of Victoria in Victoria, Canada. Prior to assuming this position, he was the founding director of the University of Victoria School of Social Work. His recent scholarly work includes an examination of preventive strategies in child welfare and of the implementation phase of the policy-making process.

INDEX